Advancing Methods for Interdisciplinarity in the Social Sciences

Advancing Methods for Interdisciplinarity in the Social Sciences is a new book series from Palgrave that supports, encourages and facilitates excellence in interdisciplinary research in the social sciences and beyond.

As social scientists are increasingly urged to turn to interdisciplinary research to tackle urgent and complex problems, both in their own work and within research teams, new practices are emerging to fit these hybrid approaches. Collaborations with STEM and humanities scholars especially both demand and inspire innovative methodological interventions.

This exciting new series presents focused volumes that cover a broad spectrum of qualitative, quantitative, and mixed research methods that can be applied across various disciplines and fields. Publications are applied and practical, as well as theoretical and reflective, including comprehensive overviews and analyses of methods to enable direct applications to a variety of socio-cultural and scientific contexts. Experience-led and applied in nature, each book in the series will demonstrate practical, ethical, and innovative approaches to facilitate dialogue and collaboration across disciplines.

Showcasing cutting-edge developments in research methods and methodological approaches, the series facilitates creative cross-pollination between disciplines to equip social scientists with future-facing tools for high quality, transformative research.

We welcome proposals from researchers interested in rethinking methodologies from interdisciplinary perspectives, for short Pivot format (25,000–50,000 words) and standard length (maximum of 80,000 words) edited and authored volumes.

Lesley Gourlay, UCL, UK
Lois Presser, University of Tennessee, USA
Michael Chataway, Queensland University of Technology, Australia
Naveen Thayyil , IIT Delhi, India
Nikki Fairchild, University of Portsmouth, UK
Phil Murphy, Middlebury Institute, USA
Roger Norum, University of Oulu, Finland
Seyyed-Abdolhamid Mirhosseini, Hong Kong University, Hong Kong
Simon Hayhoe, University of Exeter, UK
Wayne Fife, Memorial University of Newfoundland, Canada

Publisher Contacts:
Marion Duval (marion.duval@palgrave.com)
Clelia Petracca (clelia.petracca@palgrave.com)
Rebecca Longtin (rebecca.longtin@palgrave-usa.com)

Anthony Jorm

Using the Delphi Method to Establish Expert Consensus

A Practical Guide

Anthony Jorm
Melbourne School of Population & Global Health
University of Melbourne
Melbourne, VIC, Australia

This work was supported by the National Health and Medical Research Council
(Investigator Grant APP1172889).

ISSN 3059-3506   ISSN 3059-3514 (electronic)
Advancing Methods for Interdisciplinarity in the Social Sciences
ISBN 978-981-96-8356-7   ISBN 978-981-96-8357-4   (eBook)
https://doi.org/10.1007/978-981-96-8357-4

This Palgrave Macmillan imprint is published by the registered company Springer Nature
Singapore Pte Ltd.
The registered company address is: 152 Beach Road, #21-01/04 Gateway East, Singapore
189721, Singapore

If disposing of this product, please recycle the paper.

PREFACE

Writing this book on the Delphi method has led me to reflect back on how I first got to know about and use it. My earliest recollection is from the late 1970s when I was a university lecturer teaching an introductory psychology course. The course textbook we used had a short paragraph on the use of the Delphi method for forecasting the future. For many years afterwards, I thought this was the only application of the method.

In the 1990s, there was an increase in Delphi studies in health sciences, with many applications that were quite different from the original use in forecasting. My first personal involvement was around 2000, when I was asked to be an expert panellist on a Delphi study on what should be the core functions of public health practice in Australia (National Public Health Partnership, 2000). This application involved coming to a consensus on the priority areas for practice.

Later, in 2005, I was asked to be an expert panellist and co-author on a Delphi study to estimate the global prevalence of dementia (Ferri et al., 2005). The problem that this Delphi study aimed to deal with was the patchiness of prevalence data from various regions of the world. There was reasonable data from Western Europe, North America, East Asia, South Asia, and Australasia, but very limited data from regions such as Africa, Eastern Europe, the Middle East, and Latin America. As experts, we had to make the best possible estimates for each region, given the data available.

My major involvement with the Delphi method began after my wife Betty Kitchener and I developed the world's first Mental Health First Aid training course in 2000. This course aimed to train members of the public on how to assist a person who was developing a mental health problem or was experiencing a mental health crisis (e.g. they were suicidal or self-harming). In developing the content of this course, we wanted to make it as evidence-based as possible. However, we found that while there were clinical practice guidelines for professionals, there was virtually no evidence to guide what members of the public should do in a mental health first aid situation. In our first edition of the course manual (Kitchener & Jorm, 2002), we used what little evidence was available, along with advice from professional colleagues with relevant expertise and applied the principle that being caring towards people with mental health problems would be helpful.

We realized that this was not a sound enough basis for the course content and sought a way of developing mental health first aid guidelines. Because this was an area where randomized controlled trials were neither feasible nor ethical, we settled on expert consensus as the best possible evidence and Delphi as a method of achieving this. In 2006, we were successful in getting a grant from Australian Rotary Health to use the method to develop guidelines for assisting a person in a mental health crisis. When using the Delphi method for this purpose, we had to decide who were the appropriate experts on mental health first aid. No one group seemed ideal, so we set up separate panels of professionals, people with personal experience of a mental health problem or crisis, and carers of those with personal experience. To be included in our mental health first aid guidelines, a first aid action had to have at least 80% agreement as "important" or "essential" from all three panels. The guidelines were used to develop a second edition of the Mental Health First Aid Manual and course (Kitchener et al., 2010). Mental Health First Aid training has now spread to 29 other countries, with over 8 million people trained, and the strategies they are taught are based on Delphi expert consensus studies.

Subsequently, I became an investigator on many other Delphi studies on mental health topics. Besides those to develop mental health first aid guidelines, there were studies on priorities for research, self-help strategies for depressive symptoms, preventive strategies for parents, strategies for communicating with young people about sensitive topics, guidelines for journalists on reporting of crime and mental illness, suicide postvention guidelines for secondary schools, and core competencies for early psychosis clinicians. As of 2024, I have been an author of 60 published Delphi studies and 2 reviews on the Delphi method. One of these reviews on how to use the method in mental health research has been highly cited, with over 800 citations in Google Scholar (Jorm, 2015).

Despite my growing respect for the Delphi method in providing useful guidance on health actions, I was very aware that expert consensus is seen by many researchers as a poor alternative to randomized controlled trials and other types of quantitative studies. In medical research, for example, a number of organizations working in the area of evidence-based medicine have published hierarchies of evidence for evaluating treatments, in which systematic reviews of randomized trials are the highest form of evidence, and expert opinion, if it is mentioned at all, is ranked at the bottom (Burns et al., 2011). There is a certain irony here in that these hierarchies of evidence are themselves based on expert opinion. However, such views showed a need for some strong justification that the Delphi method could produce valid findings.

This gap was filled by reading James Surowiecki's (2004) popular social science book, The Wisdom of Crowds. Surowiecki presented many examples to support his argument that under certain conditions the aggregated judgements of groups of people (which he called "crowds") could produce more accurate results than individual experts. I only came across Surowiecki's book a decade after it had been published, but it led me to a rich scientific literature on the validity of group judgement. Although the Delphi method was developed many decades before wisdom-of-crowds became a popular topic of research, it can provide a much-needed empirical grounding for the validity of expert consensus and can suggest ways in which consensus processes can be improved.

Over the years of thinking about the role of expert consensus in science, I became convinced that rather than just being a low-quality methodology that could be used when superior methods were not feasible, consensus processes actually pervaded all of science. This led me to write a book on Expert Consensus in Science arguing the case for its importance in establishing scientific truth, for guiding science-based policy and professional practice, and for determining the validity of research methods (Jorm, 2025). The present book on the Delphi method is complementary to the earlier work and focuses on the most common method of formally establishing an expert consensus.

There are already a number of highly cited journal articles on how to carry out a Delphi study, including my own article on its application in mental health research. So, why write a book on the topic? Journal articles are necessarily short, thus limiting the amount of detail that can be provided. In writing a longer work on the Delphi method, my aim was to

- Take a broad cross-disciplinary approach, in contrast to previous work, which has typically covered the use of the Delphi method in one specific discipline or professional area.
- Relate the Delphi method to research on the wisdom of crowds and to use this research to suggest ways in which the validity of the method can be improved.
- Provide practical guidance on the many choices that have to be made when planning and carrying out a Delphi study.
- Discuss the challenges involved in putting Delphi findings into action. Publication of a Delphi study is unlikely in itself to produce any change in practice, so this needs to be considered in advance when a study is being planned.

A number of people have assisted with this book. My wife Betty Kitchener has been a co-investigator with me on many Delphi studies and provided very valuable comments on an earlier draft that led to major improvements in the organization of the book. The publisher also provided reviews of the book proposal and final draft, which led to a number of improvements. I thank these anonymous reviewers. Finally, I would like to thank Marion Duval, the Editor of the Palgrave Macmillan series on Advancing Methods for Interdisciplinarity in the Social Sciences, for her prompt feedback and ongoing support for this book.

Melbourne, VIC, Australia   Anthony Jorm

REFERENCES

Burns, P. B., Rohrich, R. J., & Chung, K. C. (2011). The levels of evidence and their role in evidence-based medicine. Plastic and Reconstructive Surgery, 128(1), 305–310. https://doi.org/10.1097/PRS.0b013e318219c171

Ferri, C. P., Prince, M., Brayne, C., Brodaty, H., Fratiglioni, L., Ganguli, M., Hall, K., Hasegawa, K., Hendrie, H., Huang, Y., Jorm, A., Mathers, C., Menezes, P. R., Rimmer, E., & Scazufca, M. (2005). Global prevalence of dementia: A Delphi consensus study. Lancet, 366(9503), 2112–2117. https://doi.org/10.1016/s0140-6736(05)67889-0

Jorm, A. F. (2015). Using the Delphi expert consensus method in mental health research. The Australian and New Zealand Journal of Psychiatry, 49(10), 887–897. https://doi.org/10.1177/0004867415600891

Jorm, A. (2025). Expert consensus in science. Palgrave Macmillan. https://doi.org/10.1007/978-981-97-9222-1

Kitchener, B. A., & Jorm, A. F. (2002). Mental health first aid manual. Centre for Mental Health Research.

Kitchener, B. A., Jorm, A. F., & Kelly, C. M. (2010). Mental health first aid manual (2nd ed.). Orygen Youth Health Research Centre.

National Public Health Partnership. (2000). Public health practice in Australia today: A statement of core functions. National Public Health Partnership.

Surowiecki, J. (2004). The wisdom of crowds: Why the many are smarter than the few. Doubleday.

COMPETING INTERESTS

The author has no competing interests to declare that are relevant to the content of this manuscript.

CONTENTS

LIST OF FIGURES

LIST OF TABLES

CHAPTER 1

Origins and Uses of the Delphi Method

Abstract This chapter describes the development of the Delphi method by the RAND Corporation in the 1950s, early criticisms of the method, and its subsequent evolution and growth in use. A wide range of variants of the Delphi method have been developed, only some of which aim to establish a consensus. The chapter examines contemporary uses of the method to establish consensus, covering its uses in making judgements on facts where the evidence is imperfect, setting methodological standards, making predictions, defining foundational concepts, determining collective values, improving professional practice, and improving policy.

Keywords Delphi method • Criticisms • Uses • Variants

Before describing the origin of the Delphi method, I need to discuss its pronunciation. In the English-speaking world there are two different pronunciations. In North America it is commonly pronounced "Del-fye", while in other countries "Del-fee" is more usual. In Greek, the final syllable is pronounced "fee", which could be used to argue that the latter pronunciation is correct. However, both are common in English and are best seen as a dialectical variation with both being acceptable. Speakers of other languages will have to consult their local conventions for the accepted pronunciation.

© The Author(s) 2025
A. Jorm, *Using the Delphi Method to Establish Expert Consensus,*
Advancing Methods for Interdisciplinarity in the Social Sciences,
https://doi.org/10.1007/978-981-96-8357-4_1

1

In this book, I refer to the "Delphi method". However, it is also commonly referred to as the "Delphi technique" or "Delphi process". These terms are essentially interchangeable, although the term "process" tends to place more emphasis on the steps involved in carrying out a Delphi study.

In contemporary use, there are many variations of the method, but a typical Delphi study would involve the following key steps:

- A researcher poses a question that is answerable by the method.
- The researcher recruits a group of experts on the topic.
- The researcher creates a questionnaire with statements for the experts to rate for agreement or provide estimates.
- The experts complete the questionnaire independently, and the researcher collates their responses.
- The researcher provides anonymous feedback on the responses of the group, which may be quantitative or qualitative feedback.
- Experts can revise their responses after seeing the group feedback.
- The process is repeated through multiple rounds, with some statistical criteria used to define consensus.

1   Development by the RAND Corporation

The Delphi method was developed in the 1950s by the RAND Corporation, a US nonprofit organization which was set up "To further and promote scientific, educational, and charitable purposes, all for the public welfare and security of the United States" (RAND, 2025). Much of the organization's work during this period was in areas of importance for US defence during the Cold War, so was classified and not available to the public. This included the early work using the Delphi method.

The method takes its name from a place in Greece, which in ancient times was the site of a temple with a priestess (the oracle) who could deliver prophecies from the god Apollo. The oracle was consulted before all major undertakings in the ancient Greek world, including whether to start a war or found a colony. The adoption of this name for the Delphi method reflects its initial use in forecasting.

According to one of its founders at RAND, Delphi is "a method of eliciting and refining group judgments" (Dalkey, 1969). The first Delphi study was carried out by Olaf Helmer and Norman Dalkey in 1951, with the results reported in a classified document titled *The Use of Experts for the Estimation of Bombing Requirements*. This study remained unknown to the broader research community until it was declassified a decade later (Dalkey & Helmer, 1962) and then published in the journal *Management Science* (Dalkey & Helmer, 1963). In the report, the authors state that the Delphi method "was devised in order to obtain the most reliable opinion consensus of a group of experts by subjecting them to a series of questionnaires in depth interspersed with controlled opinion feedback" (Dalkey & Helmer, 1962, p. v). The study aimed to use expert opinions to determine the best US industrial targets from the perspective of a Soviet strategic planner and to estimate how many atomic bombs would be needed to decrease munitions production by a specific amount. The expert panel consisted of seven people, four of whom were economists, one a physical-vulnerability specialist, one a systems analyst, and one an electronics engineer. The experts completed five questionnaires altogether at approximately weekly intervals, with feedback on the responses after each questionnaire. The questionnaires were designed to get estimates of the number of bombs required, to allow experts to explain their reasoning, to list causal factors, to estimate those factors, and to suggest what additional data would help them give more confident estimates. The collective results from each round of questionnaires were shared with experts along with a new set of questions and more information. The experts were then given the chance to revise their earlier answers. It was expected that the experts' estimates would gradually converge. This was found to be the case, with estimates of the number of bombs required ranging from 50 to 5000 in the first questionnaire and narrowing to 167 to 360 in the fifth questionnaire. The median response of the experts, which was 276 bombs on the final questionnaire, was taken as the consensus.

While this was the first Delphi study, it did not stimulate more widespread use of the method. The more influential study was one carried out by the RAND Corporation in the 1960s on long-range forecasting, which became the prototype for subsequent Delphi studies (Dayé, 2018). This study, by Theodore Gordon and Olaf Helmer (1964), was concerned with expert predictions of advances up to 50 years in the future in six areas: scientific breakthroughs, population control, automation, space progress, and prevention of war and weapons systems. They recruited experts in each of these six areas, inviting 150 persons and receiving responses to one or more questionnaires from 82 of these. Each panel responded to 4 questionnaires spaced around 2 months apart. Examples of the items are "Chemical control over heredity—molecular biology", "Popular use of personality control drugs", and "Reliable weather forecasts". The experts had to give an approximate year by which they thought there was a 50% chance of an advance occurring. Consensus was regarded as being achieved if a panel had a tight interquartile range in their year estimates. After each questionnaire, the panel was given feedback on the median year that was estimated and a summary of the reasons for a deviating opinion on the part of the minority. The estimates are interesting from a vantage of six decades later. Some estimates look reasonable in retrospect, such as "Effective fertility control by oral contraceptive" (median estimate 1970), but others were wildly optimistic, such as "Automatic language translators" (1972), "Creation of a primitive form of artificial life" (1989), and "Feasibility of limited weather control, in the sense of substantially affecting regional weather at reasonable cost" (1990).

The long-range forecasting study illustrates what became the three defining features of the Delphi method. According to Dalkey (1969, p. v), these are

(1) *Anonymous response*—opinions of members of the group are obtained by a formal questionnaire.

(2) *Iteration and controlled feedback*—interaction is effected by a systematic exercise conducted in several iterations, with carefully controlled feedback between rounds.

(3) *Statistical group response*—the group opinion is defined as an appropriate aggregate of individual opinions in the final round. These features are designed to minimize the biasing effects of dominant individuals, of irrelevant communications, and of group pressure towards conformity.

This study differed from the first Delphi study in having a much greater number of experts. However, this was at the expense of reducing open-ended questions, the amount of qualitative information gathered on reasons for judgements, and communication of shared data to the experts to inform their judgements. Instead, questions became more structured and the feedback more quantitative. Only experts who deviated most strongly from the median were asked to justify their opinions, limiting the exchange of reasoning among participants. Dayé (2018) has argued that this simplification of the method led to a loss of the rich exchange of reasons for judgements which provides a justification for a consensus. Whatever the merits of the two approaches, the differences between them illustrate the considerable variation in the procedure that can occur when using the Delphi method. Rather than a single monolithic method, Delphi is best thought of as a general family of techniques for establishing a group consensus, which share some common principles.

2 EARLY CRITICISMS OF THE METHOD

Although the Delphi method was developed within the RAND Corporation, it was not accepted as a suitable method of forecasting by all its employees. In 1974, RAND published a report by Harold Sackman (1974) giving a critical analysis of the method. The report had the innocuous title *Delphi Assessment: Expert Opinion, Forecasting, and Group Process*, which gives no indication of its scathing conclusions.

> The analysis concludes that conventional Delphi is basically an unreliable and scientifically unvalidated technique in principle and probably in practice. In the absence of a comprehensive survey of the extensive applications literature, it is suggested, but not proven, that the results of most Delphi experiments are probably unreliable and invalid. Even variations of conventional Delphi should not be encouraged unless they explicitly attempt to meet the challenges of generally accepted standards of rigorous empirical experimentation in the social sciences. ... The final recommendation is that conventional Delphi be dropped from institutional, corporate, and government use until its principles, methods, and fundamental applications can be established experimentally as scientifically tenable. (p. vi)

Sackman (1975) subsequently published his criticisms in a peer-reviewed journal article. Because the Delphi method used questionnaires to assess group opinion, Sackman's approach was to evaluate the method against contemporary standards for opinion polling and questionnaire design. In particular, he compared Delphi studies to the standards of the American Psychological Association's *Standards for Educational and Psychological Tests and Manuals* (American Psychological Association, 1966). He came up with 16 conclusions that were highly critical of this method. His criticisms covered questionnaire design ("Is often characterized by crude questionnaire design"), the selection of experts ("is highly vulnerable in its concept of 'expert' with unaccountable sampling and in its selection of panellists, expert or otherwise"), its selection of questionnaire items ("abdicates responsibility for item population sampling in relation to theo-retical constructs for the object area of inquiry"), its approach to assessing a consensus ("capitalizes on forced consensus based on group sugges-tion"), the lack of group discussion and debate ("denigrates group and face-to-face discussion and, without proof, claims that the use of anony-mous group opinion is superior to competing approaches"), the lack of attention to precision in statistical summaries of opinion ("gives an exag-gerated illusion of precision, misleading uninformed users of the results"), and the lack of data to support the assumptions of the method ("produces virtually no serious literature to test basic assumptions and alternative hypotheses") (Sackman, 1975, pp. 715–716).

Criticisms also appeared at the same time outside of RAND. Hill and Fowles (1975) published a critique of Delphi as a method of forecasting, focusing on issues of reliability and validity. They reviewed the evidence available at the time showing a lack of agreement in predictions across dif-ferent Delphi studies. They also argued that many of the questions used in Delphi forecasting studies were too vague to be useful for planning, that who qualified as an "expert" was not well defined, that there were poten-tial biases in the selection of future events to be estimated, that the response rates to questionnaires were low, that there was a lack of clarity on what defines "consensus", and that there were potential pressures on expert panellists to conform to the group feedback in the interests of har-mony rather than to make the effort to justify their disagreements to the group. Arguably, the most important limitation was the lack of evidence that the forecasts were valid, as indicated by the accurate prediction of what actually occurred.

Some of these criticisms relate to the specific application of forecasting, but others potentially apply to Delphi studies more broadly. However, the use of the method has evolved since the early Delphi studies and in con-temporary applications researchers have often sought to overcome the weaknesses pointed out by the early critics. Nevertheless, their criticisms point out key issues that researchers have to consider when designing Delphi studies, which are dealt with in detail in Chap. 4.

3   SUBSEQUENT GROWTH IN USE

Despite its rocky start, the Delphi method has become increasingly used. Its application has moved beyond the specific problem of forecasting for defence purposes, to be used for a wide range of purposes across many disciplines.

To mark the sixtieth anniversary of the publication of the first Delphi study in a peer-reviewed journal, Khodyakov et al. (2023) carried out a systematic literature search of peer-reviewed publications in the English language. They found 19,832 articles that used the method and a further 627 that dealt with the Delphi methodology. As shown in Fig. 1.1, the growth of Delphi studies from the 1960s to the 2010s has been spectacular. This trend has continued in the 2020s, with nearly a third of all articles ever published ($n = 6454$) appearing in the first 2.5 years of this decade.

Khodyakov et al. (2023) also examined the disciplines that used the Delphi method. They found that 65% of the articles were from medical journals, 15% from science and technology, 15% from social sciences, and 4% from other areas. The predominance of medical uses was found to have increased over time. There were no Delphi articles in medical journals in the 1960s, but this sharply increased to over two-thirds of the articles by the 2020s.

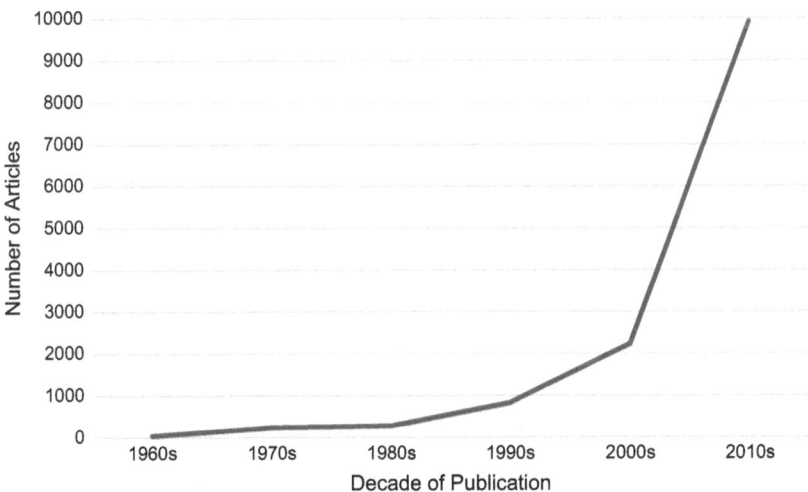

Fig. 1.1 Number of published articles that used the Delphi method per decade from the 1960s to the 2010s (based on data from Khodyakov et al., 2023)

4   Delphi Variants

There can be considerable differences between studies that are described as using a Delphi method. Some Delphi studies are described by their authors as being a "Classical Delphi". This term typically refers to the Delphi methodology that was most widely adopted after it was disseminated outside the RAND Corporation in the 1960s. Many Delphi studies involve variants of this method and are sometimes described as using a "modified Delphi". However, even from the earliest years when Delphi methods were being developed by the RAND Corporation, there were important differences in methodology from study to study (Dayé, 2018). Not all conformed to the same "Classical Delphi" method.

Some researchers have given specific names to variants of the Delphi method. These are listed together with their key features in Table 1.1. The table also shows the "Classical Delphi" as a point of contrast. Most of the named variants are seldom used. In some cases, the departures from the features of the Classical Delphi are so great that it is unfortunate that the label "Delphi" has been applied to them. Some of the methods do not aim for consensus at all (Policy Delphi, Dissensus Delphi, Argument Delphi, Disaggregative Policy Delphi). They are therefore not discussed further in this book, given that the focus here is on use of Delphi methods for establishing consensus. Others retain the aim of establishing a consensus but involve major changes in the way panellists interact or the feedback they receive. These variations are discussed later in the book as options that the researcher can consider when designing their study.

One of the reasons for the wide range of methods that have been labelled as "Delphi" is that the method has sometimes been defined so broadly that it loses any specificity of meaning. For example, Linstone and Turoff (2002, p. 3) have defined Delphi as "a method for structuring a group communication process so that the process is effective in allowing a group of individuals, as a whole, to deal with a complex problem". In this definition, there is no mention of consensus as an aim and no restrictions are imposed on how the group communicates. The definition could encompass many of the alternative methods for establishing a consensus, including expert working groups, the Nominal Group Technique, consensus conferences, and prediction markets.

Table 1.1   Named variants of the Delphi method

Variant name	Aim	Distinguishing characteristics
Classical Delphi (Dalkey, 1969)	Establish a consensus	• Anonymity of responses. • Feedback on responses to panellists followed by re-surveying. • Statistical summary of group response.
Policy Delphi (Turoff, 1970)	Establish the range of positions on an area of policy and the arguments for and against these positions. The goal is not to establish a consensus	• Requires a highly diverse panel in terms of policy positions. • The output can be used by a small committee to develop a policy.
Delphi Conferencing (Turoff, 1971)	Conduct an online conference discussion on a topic with anonymity of participants	• All interaction is online. • Discussion feedback provided in real time. • Conference is run asynchronously over several weeks.
Decision Delphi (Rauch, 1979)	Assist decision-makers to make decisions	• Panel members are all actual decision-makers. • Anonymity is not strictly observed.
Group Delphi (Webler et al., 1991)	Consolidate expert opinion in a very short time period	• No anonymity. • During a plenary session, a moderator asks experts to justify their opinion. • Discussion among the panel encouraged.
Fuzzy Delphi Method (Kaufmann & Gupta, 1988)	Establish a consensus estimate where there is ambiguity about the concept	• Asks experts to make estimates reflecting the uncertainty of the concept (e.g. minimum, most plausible and maximum estimate). • Analysis of estimates is based on fuzzy numbers.
Argument Delphi (Kuusi, 1999)	Find relevant arguments for different positions and discussing these	• Focuses on generating and criticizing arguments rather than establishing consensus. • Participants are informed about the names of other participants, but responses are given anonymously.

(continued)

Table 1.1 (continued)

Variant name	Aim	Distinguishing characteristics
Disaggregative Policy Delphi (Tapio, 2003)	Provide an improved method of feedback between survey rounds in a Policy Delphi	• Cluster analysis is used to group quantitative responses and combined with qualitative arguments.
Real-time (RT) Delphi (Gordon & Pease, 2006)	Shorten the time required for Delphi studies by collecting responses in real time without rounds	• Questions are answered online. • After logging in, a participant receives a summary of previous responses and reasons for answers. • The participant provides their response immediately after receiving the summary.
Dissensus Delphi (Steinert, 2009)	Generate expert opinion on a topic through encouraging dissensus	• Forces experts to give varying opinions from the ones already inputted. • Experts selected to be as heterogeneous as possible. • A facilitator synthesizes opinions in a report.
Grounded Delphi (Päivärinta et al., 2011)	Create theory	• Brainstorming is used to collect qualitative data from experts. • Grounded Theory methodology used to develop concepts from the qualitative data. • Delphi method used to establish consensus on core concepts.
Delphi Market (Prokesch et al., 2015)	Make better predictions by combining the Delphi method with prediction markets	• Used for a forecasting question that is verifiable. • Uses experts on a topic in a prediction market.

(*continued*)

Table 1.1 (continued)

Variant name	Aim	Distinguishing characteristics
IDEA Protocol (Hemming et al., 2018)	Estimate quantities or probabilities and their uncertainties	• Involves four steps: "Investigate", "Discuss", "Estimate", "Aggregate". • "Investigate": experts individually provide estimates and provide reasons for judgements. • "Discuss": experts receive anonymous feedback on answers and a summary of results. • "Estimate": Experts make second private estimate. • "Aggregate": Statistical summary of estimates and uncertainties calculated, which receive final review from experts.

5   Contemporary Uses to Establish Consensus

While not all studies that use the term "Delphi" aim to establish consensus, the vast majority do. To show the scope of the Delphi method's contemporary uses to establish consensus, I did a scan of Delphi publications from the 2020s listed in the Scopus database and classified them into types of uses. Table 1.2 gives the resulting taxonomy of contemporary uses together with specific examples of each type of use. I would not claim that this taxonomy covers every use that has been made of the Delphi method, but it does arguably cover the major uses. It may be useful to potential users of the method who want to decide whether it would suit the aims of their research. If a researcher's aims fit one of the broad categories of use in Table 1.2, then the Delphi method may be a suitable option.

When using Table 1.2 as a framework for potential uses, the reader should be aware that the categories are fuzzy at the edges and overlap with each other. For example, I have placed "Setting research priorities" under the broad category of "Determining collective values" on the basis that judgements about priorities reflect the values of the expert panel about what research is most important. However, setting research priorities also requires knowledge of the existing research and where there are gaps, so includes an element of "Making judgements on facts where the evidence is imperfect". Similarly, I have classified "Developing practice guidelines"

Table 1.2 Contemporary uses of the Delphi method for establishing consensus

Broad category of use	Specific uses	Examples from the literature
Making judgements on facts where the evidence is imperfect	Estimating quantities	Estimation of the number of people living with dementia at different stages of the condition in India (Farina et al., 2024)
	Determining causes and risk factors	Determining impact of screen use on sleep health across the lifespan (Hartstein et al., 2024)
Setting methodological standards	Improving the quality of research methods	Guidelines for preclinical research on potentized preparations manufactured according to current pharmacopoeias (Tournier et al., 2024)
	Improving the quality of scientific reporting	Standards for reporting of systematic reviews of outcome measurement instruments (Elsman et al., 2024)
	Determining outcome measures	Development of a core outcome set for trials in surgical handover (Ryan et al., 2024)
Making predictions	Forecasting the future	Future developments in complex health problems of older people in the Netherlands in 2040 (Baädoudi et al., 2023)
	Identifying risks	Identifying the Australian terrestrial snake and lizard species most at risk of extinction (Geyle et al., 2021)
Defining foundational concepts	Defining a concept	Defining well-being in psoriasis (Daudén et al., 2024)
	Standardizing terminology	Standardizing definitions of various fasting approaches in humans (Koppold et al., 2024)
	Developing diagnostic criteria	Diagnostic criteria for LUMBAR syndrome (Metry et al., 2024)

Determining collective values	Setting ethical standards	Revising the Taiwan code of ethics for nurses (Lu et al., 2024)
	Setting research priorities	Research priorities of members of the British Association for Surgery of the Knee (Ahmed & Metcalfe, 2024)
	Setting policy priorities	Prioritization of vaccines for introduction in the national immunization programme in the Republic of Korea (Choi et al., 2024)
	Gathering cultural knowledge	Indigenous-led modifications to improve the accessibility and usefulness of mutual support groups for substance use (Dale et al., 2021)
Improving professional practice	Determining professional competencies	Competencies required by sonographers teaching ultrasound to other professions (Cormack et al., 2024)
	Developing practice guidelines	Appropriate use of high importance-rated antimicrobials in animals (Sri et al., 2024)
	Developing a training curriculum	Developing mental health curricula and a service provision model for clinical associates in South Africa (Moodley et al., 2024)
	Developing an intervention	Developing a vape shop-based smoking cessation intervention (Langley et al., 2024)
	Developing an assessment tool	Development of a risk assessment tool for cancer-related venous thrombosis (Qin et al., 2024)
Improving policy	Developing a policy	Determining key criteria for sustainable game management as part of land use policy (Linares et al., 2024)
	Developing an economic model	Developing a fall prevention intervention economic model (Saunders et al., 2023)

under the broad category of "Improving professional practice", but it also involves a knowledge of the evidence on the effectiveness of various professional interventions and hence could also fit under "Making judgements on facts where the evidence is imperfect".

6   THE CONTRIBUTION OF THIS BOOK

The Delphi method was originally developed for the specific purpose of forecasting and was the subject of some severe criticisms in its early years. Why, then, given this bumpy start has its use expanded so widely across many fields and for a range of different purposes? An answer to this question can be found in my book, *Expert Consensus in Science* (Jorm, 2025). It argues that expert consensus plays an essential role in science and science-based practice and policy. Because of this important role, there will inevitably be a demand for methods that have the potential to validly assess consensus.

Delphi is one of a range of methods that have been used to establish consensus. Chapter 2 gives an overview of the most commonly used methods and looks at their similarities and differences from the Delphi method. This chapter may be useful to researchers who are deciding whether Delphi or another consensus method is most appropriate for their aims.

A key issue in consensus methods is whether they are likely to produce valid conclusions. Chapter 3 examines what wisdom-of-crowds research tells us about the conditions under which a group of experts is more likely to come to a valid consensus. It argues that the Delphi method comes closer to meeting these conditions than some other consensus methods. Moreover, the findings from wisdom-of-crowds research can be used to inform the design of better-quality Delphi studies.

Chapter 4 deals with the details of how to carry out a Delphi study and is consequently the longest in the book. Readers who simply want to know how to do a study and are not interested in alternative methods of consensus or the validity of consensus processes should go directly to this chapter. The chapter takes the reader through each step from framing a research aim through to reporting the results, giving examples at each step drawn from the literature.

Publication of a Delphi study is unlikely in itself to lead to a change in practice or policy. The final chapter (Chap. 5) deals with the implementation of the Delphi findings and the need to plan this from the very beginning. It also gives a range of examples of where the consensus from Delphi studies has had an impact on practice or policy.

REFERENCES

Ahmed, I., & Metcalfe, A. (2024). Research priorities of members of the British Association for Surgery of the Knee. *Bone and Joint Journal, 106* B(7), 662–668. https://doi.org/10.1302/0301-620X.106B7.BJJ-2023-0691.R1

American Psychological Association. (1966). *Standards for educational and psychological tests and manuals.* APA.

Baâdoudi, F., Picavet, S. H. S. J., Hildrink, H. B. M., Hendrikx, R., Rijken, M., & de Bruin, S. R. (2023). Are older people worse off in 2040 regarding health and resources to deal with it? – Future developments in complex health problems and in the availability of resources to manage health problems in the Netherlands [Article]. *Frontiers in Public Health, 11,* Article 942526. https://doi.org/10.3389/fpubh.2023.942526

Choi, W. S., Sung, Y., Kim, J., Seok, H., Choe, Y. J., Cheong, C., Cho, J., Lee, D. W., Shin, J. Y., & Yu, S. Y. (2024). Prioritization of vaccines for introduction in the national immunization program in the Republic of Korea. *Vaccine, 12*(8), Article 886. https://doi.org/10.3390/vaccines12080886

Cormack, C. J., Childs, J., & Kent, F. (2024). Competencies required by sonographers teaching ultrasound interprofessionally: A Delphi consensus study. *BMC Medical Education, 24*(1), Article 970. https://doi.org/10.1186/s12909-024-05933-x

Dale, E., Conigrave, K. M., Kelly, P. J., Ivers, R., Clapham, K., & Lee, K. S. K. (2021). A Delphi yarn: Applying Indigenous knowledges to enhance the cultural utility of SMART Recovery Australia [Article]. *Addiction Science & Clinical Practice, 16*(1), Article 2. https://doi.org/10.1186/s13722-020-00212-8

Dalkey, N. (1969). *The Delphi method: An experimental study of group opinion.* RAND.

Dalkey, N., & Helmer, O. (1962). *An experimental application of the Delphi method to the use of experts.*

Dalkey, N., & Helmer, O. (1963). An experimental application of the DELPHI method to the use of experts. *Management Science, 9*(3), 458–467. https://doi.org/10.1287/mnsc.9.3.458

Daudén, E., Belinchón, I., Colominas-González, E., Coto, P., de la Cueva, P., Gallardo, F., Poveda, J. L., Ramírez, E., Ros, S., Ruíz-Villaverde, R., Comellas, M., & Lizán, L. (2024). Defining well-being in psoriasis: A Delphi consensus among healthcare professionals and patients. *Scientific Reports, 14*(1), Article 14519. https://doi.org/10.1038/s41598-024-64738-6

Dayé, C. (2018). How to train your oracle: The Delphi method and its turbulent youth in operations research and the policy sciences. *Social Studies of Science, 48*(6), 846–868. https://doi.org/10.1177/0306312718798497

Elsman, E. B. M., Mokkink, L. B., Terwee, C. B., Beaton, D., Gagnier, J. J., Tricco, A. C., Baba, A., Butcher, N. J., Smith, M., Hofstetter, C., Aiyegbusi, O. L., Berardi, A., Farmer, J., Haywood, K. L., Krause, K. R., Markham, S., Mayo-Wilson, E., Mehdipour, A., Ricketts, J., et al. (2024). Guideline for reporting systematic reviews of outcome measurement instruments (OMIs): PRISMA-COSMIN for OMIs 2024. *Journal of Patient-Reported Outcomes, 8*(1), Article 64. https://doi.org/10.1186/s41687-024-00727-7

Farina, N., Rajagopalan, J., Alladi, S., Ibnidris, A., Ferri, C. P., Knapp, M., & Comas-Herrera, A. (2024). Estimating the number of people living with dementia at different stages of the condition in India: A Delphi process. *Dementia, 23*(3), 438–451. https://doi.org/10.1177/14713012231181627

Geyle, H. M., Tingley, R., Amey, A. P., Cogger, H., Couper, P. J., Cowan, M., Craig, M. D., Doughty, P., Driscoll, D. A., Ellis, R. J., Emery, J. P., Fenner, A., Gardner, M. G., Garnett, S. T., Gillespie, G. R., Greenlees, M. J., Hoskin, C. J., Keogh, J. S., Lloyd, R., et al. (2021). Reptiles on the brink: Identifying the Australian terrestrial snake and lizard species most at risk of extinction. *Pacific Conservation Biology, 27*(1), 3–12. https://doi.org/10.1071/PC20033

Gordon, T. J., & Helmer, O. (1964). *Report on a long-range forecasting study.* RAND.

Gordon, T., & Pease, A. (2006). RT Delphi: An efficient, "round-less" almost real time Delphi method. *Technological Forecasting and Social Change, 73*(4), 321–333. https://doi.org/10.1016/j.techfore.2005.09.005

Hartstein, L. E., Mathew, G. M., Reichenberger, D. A., Rodriguez, I., Allen, N., Chang, A. M., Chaput, J. P., Christakis, D. A., Garrison, M., Gooley, J. J., Koos, J. A., Van Den Bulck, J., Woods, H., Zeitzer, J. M., Dzierzewski, J. M., & Hale, L. (2024). The impact of screen use on sleep health across the lifespan: A National Sleep Foundation consensus statement. *Sleep Health, 10*(4), 373–384. https://doi.org/10.1016/j.sleh.2024.05.001

Hemming, V., Burgman, M. A., Hanea, A. M., McBride, M. F., & Wintle, B. C. (2018). A practical guide to structured expert elicitation using the IDEA protocol. *Methods in Ecology and Evolution, 9*(1), 169–180. https://doi.org/1 0.1111/2041-210X.12857

Hill, K. Q., & Fowles, J. (1975). The methodological worth of the Delphi forecasting technique. *Technological Forecasting and Social Change, 7*(2), 179–192. https://doi.org/10.1016/0040-1625(75)90057-8

Jorm, A. (2025). *Expert consensus in science.* Palgrave Macmillan. https://doi.org/10.1007/978-981-97-9222-1

Kaufmann, A., & Gupta, M. M. (1988). *Fuzzy mathematical models in engineering and management science.* Elsevier.

Khodyakov, D., Grant, S., Kroger, J., Gadwah-Meaden, C., Motala, A., & Larkin, J. (2023). Disciplinary trends in the use of the Delphi method: A bibliometric analysis. *PLoS One, 18*(8), e0289009. https://doi.org/10.1371/journal.pone.0289009

Koppold, D. A., Breinlinger, C., Hanslian, E., Kessler, C., Cramer, H., Khokhar, A. R., Peterson, C. M., Tinsley, G., Vernieri, C., Bloomer, R. J., Boschmann, M., Bragazzi, N. L., Brandhorst, S., Gabel, K., Goldhamer, A. C., Grajower, M. M., Harvie, M., Heilbronn, L., Horne, B. D., et al. (2024). International consensus on fasting terminology. *Cell Metabolism, 36*(8), 1779–1794.e1774. https://doi.org/10.1016/j.cmet.2024.06.013

Kuusi, O. (1999). *Expertise in the future use of generic technologies: Epistemic and methodological considerations concerning Delphi studies.* Government Institute for Economic Research.

Langley, T., Young, E., Hunter, A., & Bains, M. (2024). Developing a vape shop-based smoking cessation intervention: A Delphi study. *Nicotine and Tobacco Research, 26*(10), 1362–1369. https://doi.org/10.1093/ntr/ntae105

Linares, O., Martínez-Jauregui, M., Carranza, J., & Soliño, M. (2024). Bridging sustainable game management into land use policy: From principles to practice [Article]. *Land Use Policy, 145*, Article 107269. https://doi.org/10.1016/j.landusepol.2024.107269

Linstone, H. A., & Turoff, M. (2002). *The Delphi method: Techniques and applications.* Addison-Wesley Publishing Company, Advanced Book Program. https://books.google.com.au/books?id=uZ0RkAEACAAJ

Lu, M. S., Cheng, C. C., Lin, C. F., Yang, C. M., & Liao, M. N. (2024). Revising the Taiwan code of ethics for nurses. *The Journal of Nursing, 71*(4), 32–43. https://doi.org/10.6224/JN.202408_71(4).06

Metry, D., Copp, H. L., Rialon, K. L., Iacobas, I., Baselga, E., Dobyns, W. B., Drolet, B., Frieden, I. J., Garzon, M., Haggstrom, A., Hanson, D., Hollenbach, L., Keppler-Noreuil, K. M., Maheshwari, M., Siegel, D. H., Waseem, S., & Dias, M. (2024). Delphi consensus on diagnostic criteria for LUMBAR syndrome. *Journal of Pediatrics, 272*, 114101. https://doi.org/10.1016/j.jpeds.2024.114101

Moodley, S. V., Wolvaardt, J., & Grobler, C. (2024). Developing mental health curricula and a service provision model for clinical associates in South Africa: A Delphi survey of family physicians and psychiatrists. *BMC Medical Education, 24*(1), Article 669. https://doi.org/10.1186/s12909-024-05637-2

Päivärinta, T., Pekkola, S., & Moe, C. (2011). Grounding theory from Delphi studies. In *International conference on information systems 2011, ICIS 2011.*

Prokesch, T., von der Gracht, H. A., & Wohlenberg, H. (2015). Integrating prediction market and Delphi methodology into a foresight support system— Insights from an online game. *Technological Forecasting and Social Change, 97*, 47–64. https://doi.org/10.1016/j.techfore.2014.02.021

Qin, X., Gao, X., Yang, Y., Ou, S., Luo, J., Wei, H., & Jiang, Q. (2024). Developing a risk assessment tool for cancer-related venous thrombosis in China: A modified Delphi-analytic hierarchy process study. *BMC Cancer, 24*(1), Article 120. https://doi.org/10.1186/s12885-024-11877-8

RAND. (2025). *Our history*. RAND. Retrieved February 8, 2025, from https://www.rand.org/about/history.html#:~:text=RAND%27s%20articles%20of%20incorporation%20commit,contract%20for%20Project%20RAND%2C%201946

Rauch, W. (1979). The decision Delphi. *Technological Forecasting and Social Change, 15*(3), 159–169. https://doi.org/10.1016/0040-1625(79)90011-8

Ryan, J. M., Devane, D., Simiceva, A., Eppich, W., Kavanagh, D. O., Cullen, C., Hogan, A. M., & McNamara, D. A. (2024). Surgical Handover Core Outcome Measures (SH-CORE): A protocol for the development of a core outcome set for trials in surgical handover [Article]. *Trials, 25*(1), Article 373. https://doi.org/10.1186/s13063-024-08201-x

Sackman, H. (1974). *Delphi assessment: Expert opinion, forecasting, and group process*. RAND.

Sackman, H. (1975). Summary evaluation of Delphi. *Policy Analysis, 1*(4), 693–718. http://www.jstor.org/stable/42784280

Saunders, H., Anderson, C., Feldman, F., Holroyd-Leduc, J., Jain, R., Liu, B., Macaulay, S., Marr, S., Silvius, J., Weldon, J., Bayoumi, A. M., Straus, S. E., Tricco, A. C., & Isaranuwatchai, W. (2023). Developing a fall prevention intervention economic model. *PLoS One, 18*(1 January), Article e0280572. https://doi.org/10.1371/journal.pone.0280572

Sri, A., Bailey, K. E., Scarborough, R., Gilkerson, J. R., Thursky, K., Browning, G. F., & Hardefeldt, L. Y. (2024). Reaching consensus amongst international experts on the use of high importance-rated antimicrobials in animals – A Delphi study. *One Health, 19*, Article 100883. https://doi.org/10.1016/j.onehlt.2024.100883

Steinert, M. (2009). A dissensus based online Delphi approach: An explorative research tool. *Technological Forecasting and Social Change, 76*(3), 291–300. https://doi.org/10.1016/j.techfore.2008.10.006

Tapio, P. (2003). Disaggregative policy Delphi: Using cluster analysis as a tool for systematic scenario formation. *Technological Forecasting and Social Change, 70*(1), 83–101. https://doi.org/10.1016/S0040-1625(01)00177-9

Tournier, A. L., Bonamin, L. V., Buchheim-Schmidt, S., Cartwright, S., Dombrowsky, C., Doesburg, P., Holandino, C., Kokornaczyk, M. O., van de Kraats, E. B., López-Carvallo, J. A., Nandy, P., Mazón-Suástegui, J. M., Mirzajani, F., Poitevin, B., Scherr, C., Thieves, K., Würtenberger, S., & Baumgartner, S. (2024). Scientific guidelines for preclinical research on potentised preparations manufactured according to current pharmacopoeias—The PrePoP guidelines. *Journal of Integrative Medicine, 22*(5), 533–544. https://doi.org/10.1016/j.joim.2024.06.005

Turoff, M. (1970). The design of a policy Delphi. *Technological Forecasting and Social Change, 2*(2), 149–171. https://doi.org/10.1016/0040-1625(70)90161-7

Turoff, M. (1971). Delphi conferencing: Computer-based conferencing with anonymity. *Technological Forecasting and Social Change, 3*, 159–204. https://doi.org/10.1016/S0040-1625(71)80012-4

Webler, T., Levine, D., Rakel, H., & Renn, O. (1991). A novel approach to reducing uncertainty: The group Delphi. *Technological Forecasting and Social Change, 39*(3), 253–263. https://doi.org/10.1016/0040-1625(91)90040-M

CHAPTER 2

Comparison of Delphi with Other Consensus Methods

Abstract This chapter gives an overview of the most commonly used methods for establishing consensus and looks at their similarities and differences from the Delphi method. These methods are expert working groups, consensus conferences, the Nominal Group Technique, surveys of experts, prediction markets, and various bespoke consensus methods.

Keywords Consensus • Delphi method • Expert working groups • Consensus conferences • Nominal group technique • Surveys of experts • Prediction markets

The Delphi method is only one way of determining consensus. When a researcher is faced with the need for a consensus process, it can be helpful to consider the other commonly used alternatives and how these differ in key features from the Delphi method. Below is a brief outline of the major alternatives: Expert Working Groups, Consensus Conferences, the Nominal Group Technique, Surveys of Experts, Prediction Markets, and Bespoke Consensus Methods. The features of these methods can be compared with the key elements of the Delphi method. To reiterate, these are as follows:

- Anonymous responses: Group members share their opinions independently through a formal questionnaire.
- Multiple rounds and feedback: Group members receive anonymous feedback on group responses and are given the opportunity to revise their responses.
- Statistical summary of group response: Some statistical criteria are used to define consensus.

1   EXPERT WORKING GROUPS

Expert working groups are often used by scientific and professional organizations to produce consensus statements on various topics. There is no single approach for how these groups operate, but they typically identify issues, choose experts, review evidence, and reach a consensus through discussion. Examples include clinical practice guidelines for psychiatrists on the treatment of anxiety disorders (Andrews et al., 2018), ISO standards for physical measurements and laboratory equipment (ISO, 2023), and the Queensland scientific consensus on the Great Barrier Reef (State of Queensland, 2017).

2   CONSENSUS CONFERENCES

In the 1970s, the US National Institutes of Health (NIH) started using consensus conferences to address controversial medical issues. Before being discontinued in 2013, these conferences produced over 160 statements on topics like lactose intolerance and Alzheimer's prevention (National Institutes of Health, 2023). The process involved bringing together a panel of 9–16 members tasked with creating a consensus statement (Ferguson, 1996). During these conferences, the panel held sessions open to the public where invited experts presented data. Following these presentations, the panel drafted a consensus statement in private, which was then shared with the interested public in a plenary session for discussion. If needed, the panel revised the statement before it was officially adopted as the conference's outcome. While the NIH has since ceased its consensus conferences, similar processes continue to be used by various organizations around the world.

3   NOMINAL GROUP TECHNIQUE

The Nominal Group Technique (NGT) is a structured group process similar to the Delphi method, aimed at identifying problems, generating solutions, and making decisions (Harb et al., 2021). In NGT, a facilitator guides a meeting where participants first write down their ideas on a given question. These ideas are shared without discussion, followed by a group discussion, and then members independently rank or rate the ideas to reach a consensus. NGT has had various adaptations regarding how group members are selected, idea elicitation, and evaluation methods. It is less commonly used than the Delphi method for achieving scientific consensus due to its requirement for face-to-face or online meetings, in contrast to Delphi's asynchronous surveys which are generally completed online. Additionally, NGT is more suitable for a smaller number of decisions compared to Delphi. Examples of the use of NGT include determining consensus on the diagnosis of irritable bowel syndrome (Rubin et al., 2006) and finding messages suitable for use in campaigns to reduce the stigma of mental illness (Clement et al., 2010).

4   SURVEYS OF EXPERTS

Formal surveys are sometimes used to gauge the consensus among members of scientific and professional societies. These surveys are similar to the first round of a Delphi study but don't provide feedback to the participants and don't involve multiple rounds of voting. Some surveys simply report the percentage agreement on a question, whereas others have a cutoff for defining consensus. They have been widely used in fields like human-induced climate change, showing strong support for the existence of a scientific consensus (Cook et al., 2016). Other examples include polling members of the International Astronomical Union to decide on Pluto's status as a planet and the American Psychiatric Association's decision to remove homosexuality from the list of mental disorders (Zachar & Kendler, 2012). A related method is to survey authors of peer-reviewed research on a specific topic. For instance, surveys of researchers on Neanderthals showed agreement on the role of demographic factors in their extinction, though opinions varied on other causes (Vaesen et al., 2021).

5   PREDICTION MARKETS

Prediction markets let participants bet on the outcomes of future events. They have been suggested as a tool for gathering consensus predictions in science, particularly for events that will be proven right or wrong over time (Pfeiffer & Almenberg, 2010). For example, a market might be designed to pay $1 if an event occurs and $0 if it does not. Often the money amounts are hypothetical. The predictions can be traded in the market and the prices reflect the collective probability of an event happening. Prediction markets have been used successfully to predict the replicability of experiments in psychology and economics (Camerer et al., 2016; Dreber et al., 2015) and to forecast disease trends (Li et al., 2016). While useful for predictions, markets are not suitable for other consensus purposes, such as coming to an agreement on values or improving professional practice or public policy.

6   BESPOKE CONSENSUS METHODS

Some organizations, like the Intergovernmental Panel on Climate Change (IPCC), have developed complex methods for building scientific consensus (Harris, 2021). The IPCC brings together global experts to draft reports on climate change, incorporating all valid perspectives and noting any conflicting evidence. These reports undergo review by both scientists and government representatives, who must agree on every part.

Another example is The Lancet Commissions set up by the medical journal *The Lancet*, which use international panels of experts to make recommendations for health policy and practice (Lancet, 2022).

7   COMPARISON WITH THE DELPHI METHOD

These alternative methods for establishing a consensus can be compared with the key elements of the Delphi method, as shown in Table 2.1. It can be seen in the table that two of the other methods, NGT and Prediction markets, share the same key elements as the Delphi method. For these methods, other important differences from the Delphi method are also noted.

Table 2.1   Comparison of the Delphi method with other consensus methods in terms of key elements and differences

Consensus method	Anonymous responses	Multiple rounds and feedback	Statistical summary of group response	Other important differences from Delphi
Delphi	Yes	Yes	Yes	
Expert working groups	No	Possibly	No	
Consensus conferences	No	Yes	No	
Nominal Group Technique (NGT)	Yes	Yes	Yes	• Requires face-to-face or online meetings. • Only suitable for a smaller number of decisions.
Surveys of experts	Yes	No	Yes	
Prediction markets	Yes	Yes	Yes	• Only suitable for making predictions. • Not suitable for agreement on values or improving practice or policy.
Bespoke consensus methods	Possibly	Possibly	Possibly	

8   CONCLUSION

This chapter has described the methods that are commonly used as alternatives to the Delphi method for establishing a consensus and given examples of their use. It has also compared the key elements of these alternative methods with those of the Delphi method. Whether or not these differences between Delphi and other methods are important for good-quality consensus judgements is the subject of the next chapter.

REFERENCES

Andrews, G., Bell, C., Boyce, P., Gale, C., Lampe, L., Marwat, O., Rapee, R., & Wilkins, G. (2018). Royal Australian and New Zealand College of Psychiatrists clinical practice guidelines for the treatment of panic disorder, social anxiety disorder and generalised anxiety disorder. *Australian and New Zealand Journal of Psychiatry, 52*, 1109–1172. https://doi.org/10.1177/0004867418799453

Camerer, C. F., Dreber, A., Forsell, E., Ho, T. H., Huber, J., Johannesson, M., Kirchler, M., Almenberg, J., Altmejd, A., Chan, T., Heikensten, E., Holzmeister, F., Imai, T., Isaksson, S., Nave, G., Pfeiffer, T., Razen, M., & Wu, H. (2016). Evaluating replicability of laboratory experiments in economics. *Science, 351*, 1433–1436. https://doi.org/10.1126/science.aaf0

Clement, S., Jarrett, M., Henderson, C., & Thornicroft, G. (2010). Messages to use in population-level campaigns to reduce mental health-related stigma: Consensus development study. *Epidemiologia e Psichiatria Sociale, 19*(1), 72–79. https://doi.org/10.1017/s1121189x00001627

Cook, J., Oreskes, N., Doran, P. T., Anderegg, W. R. L., Verheggen, B., Maibach, E. W., Carlton, J. S., Lewandowsky, S., Skuce, A. G., Green, S. A., Nuccitelli, D., Jacobs, P., Richardson, M., Winkler, B., Painting, R., & Rice, K. (2016). Consensus on consensus: A synthesis of consensus estimates on human-caused global warming. *Environmental Research Letters, 11*(4), 048002. https://doi.org/10.1088/1748-9326/11/4/048002

Dreber, A., Pfeiffer, T., Almenberg, J., Isaksson, S., Wilson, B., Chen, Y., Nosek, B. A., & Johannesson, M. (2015). Using prediction markets to estimate the reproducibility of scientific research. *PNAS, 112*, 15343–15347. https://doi.org/10.1073/pnas.1516179112

Ferguson, J. H. (1996). The NIH Consensus Development Program. The evolution of guidelines. *International Journal of Technology Assessment in Health Care, 12*(3), 460–474. https://www.ncbi.nlm.nih.gov/pubmed/8840666

Harb, S. I., Tao, L., Peláez, S., Boruff, J., Rice, D. B., & Shrier, I. (2021). Methodological options of the nominal group technique for survey item elicitation in health research: A scoping review. *Journal of Clinical Epidemiology, 139*, 140–148. https://doi.org/10.1016/j.jclinepi.2021.08.008

Harris, R. (2021, June 29). Climate explained: How the IPCC reaches scientific consensus on climate change. *The Conversation.* https://theconversation.com/climate-explained-how-the-ipcc-reaches-scientific-consensus-on-climate-change-162600

ISO. (2023). *ISO/TC 48 laboratory equipment.* International Organization for Standardization. Retrieved February 19, 2023, from https://www.iso.org/committee/48908.html

Lancet. (2022). *Information for authors.* www.thelancet.com

Li, E. Y., Tung, C. Y., & Chang, S. H. (2016). The wisdom of crowds in action: Forecasting epidemic diseases with a web-based prediction market system. *International Journal of Medical Informatics, 92*, 35–43. https://doi.org/10.1016/j.ijmedinf.2016.04.014

National Institutes of Health. (2023). *NIH consensus development program.* Retrieved February 21, 2023, from https://consensus.nih.gov/

Pfeiffer, T., & Almenberg, J. (2010). Prediction markets and their potential role in biomedical research – A review. *Biosystems, 102*(2–3), 71–76. https://doi.org/10.1016/j.biosystems.2010.09.005

Rubin, G., De Wit, N., Meineche-Schmidt, V., Seifert, B., Hall, N., & Hungin, P. (2006). The diagnosis of IBS in primary care: Consensus development using nominal group technique. *Family Practice, 23*(6), 687–692. https://doi.org/10.1093/fampra/cml050

State of Queensland. (2017). *2017 consensus statement: Land use impacts on Great Barrier Reef water quality and ecosystem condition.* State of Queensland. https://researchonline.jcu.edu.au/50116/1/2017-scientific-consensus-statement-summary.pdf

Vaesen, K., Dusseldorp, G. L., & Brandt, M. J. (2021). An emerging consensus in palaeoanthropology: Demography was the main factor responsible for the disappearance of Neanderthals. *Scientific Reports, 11*(1), 4925. https://doi.org/10.1038/s41598-021-84410-7

Zachar, P., & Kendler, K. S. (2012). The removal of Pluto from the class of planets and homosexuality from the class of psychiatric disorders: A comparison. *Philosophy, Ethics, and Humanities in Medicine, 7*, 4. https://doi.org/10.1186/1747-5341-7-4

CHAPTER 3

Conditions for a Valid Consensus

Abstract This chapter examines what wisdom-of-crowds research tells us about the conditions under which a group of experts is more likely to come to a valid consensus. It summarizes four conditions for a wise crowd: selection for expertise, cognitive diversity, independence of judgements, and opportunity for sharing. It argues that the Delphi method comes closer to meeting these conditions than some other consensus methods and that the findings from wisdom-of-crowds research can be used to inform the design of better-quality Delphi studies.

Keywords Consensus • Wisdom of crowds • Expertise • Cognitive diversity • Independence of judgements • Sharing of judgements • Delphi method

A critical issue for all consensus methods is whether they produce valid judgements. However, in most applications of consensus methods, there is no independent correct answer that can be used to assess the validity. The early developers of the Delphi method recognized this and attempted to validate the method in a series of experiments in which university students were asked to answer general knowledge questions where a correct answer was available (e.g. "What percent of the households in the US had telephones in 1965?") or to make predictions about events that could be

© The Author(s) 2025
A. Jorm, *Using the Delphi Method to Establish Expert Consensus*,
Advancing Methods for Interdisciplinarity in the Social Sciences,
https://doi.org/10.1007/978-981-96-8357-4_3

evaluated within a short follow-up time (e.g. "What will be the highest temperature recorded during June in California?") (Brown et al., 1969; Dalkey, 1969; Dalkey & Brown, 1971). However, in general, there has been very little research on the validity of the consensus methods reviewed in Chap. 2, which is surprising given how commonly they are used.

Fortunately, there has been a related area of research that has examined the conditions under which groups make better-quality judgements. Although this research has been largely independent of Delphi and other consensus methods, it can provide useful information to test their validity and to improve consensus methods. This area of research is often called the "wisdom of crowds" or "collective intelligence". I use the former term here.

1   THE WISDOM OF CROWDS

The term "wisdom of crowds" comes from a popular social science book by James Surowiecki (2004), which argued that the aggregated judgements of groups of people (a "crowd") with imperfect expertise could be surprisingly accurate and often better than the judgements of the best individual experts. Subsequently, many researchers on collective judgements adopted Surowiecki's term.

The earliest evidence for the wisdom of crowds came from Francis Galton (1907), who reported on the data from a guessing competition at an English livestock show. People visiting the show could pay a small fee to enter the competition, which involved guessing the weight of an ox after it had been slaughtered and dressed. Galton analysed the distribution of guesses, which ranged from 1074 to 1293 pounds, and found that the median estimate of 1207 pounds was close to the actual weight of 1198 pounds. The explanation for this impressive accuracy is that the errors of estimation of individuals cancelled each other out, so that the central tendency of the distribution provided a better estimate than that of most individuals.

More modern research confirms the observation that aggregated crowd judgements can perform well. I will mention two examples to illustrate this. This first example is a study of how well an online crowd and soccer experts predicted the outcomes of the 2014 World Cup matches (O'Leary, 2017). Participants in Yahoo's "World Soccer Pick'em" guessed winners (or draws) and scores, while Yahoo also gathered predictions from five soccer experts. The crowd's majority predictions were correct for 45 out of 64 matches, outperforming the experts whose correct predictions ranged from 33 to 40.

The second example concerns fact-checking news headlines. Expert fact-checking can be costly, but the combined judgements of laypeople offer a cheaper alternative. Crowds of fewer than 20 people have been shown to match the accuracy of professional fact-checkers in identifying misleading news (Martel et al., 2024). In one study, social media users from various countries rated the accuracy of COVID-19 news headlines. Researchers found that data from just 15 raters could correctly distinguish true from false headlines over 90% of the time, using fact-checking websites as a benchmark (Arechar et al., 2023).

2   THE FOUR CONDITIONS FOR A WISE CROWD

While crowds can often make wise judgements, this is not always the case. However, better-quality judgements are more likely under certain conditions. I have previously proposed four such conditions: Selection for Expertise, Cognitive Diversity, Independence of Judgements, and Opportunity for Sharing (Jorm, 2025). These are important to consider when designing and evaluating Delphi and other consensus studies.

2.1   Selection for Expertise

The wisdom-of-crowds effect relies on combining judgements from individuals who may have limited expertise. In contrast, Delphi and other consensus studies typically involve highly knowledgeable individuals. It might seem obvious that crowds with higher expertise outperform those with lower expertise, and research confirms this. The findings from two large studies are described below to illustrate this point.

Mannes et al. (2014) analysed 90 datasets to evaluate the accuracy of individual and group judgments. Forty datasets came from lab experiments where participants made numerical estimates (e.g. temperatures, distances) with group sizes ranging from 15 to 413. The remaining 50 datasets involved professional economists forecasting economic indicators (e.g. GDP, consumer price index), with a median group size of 35. The researchers compared three groups: the entire crowd; a selected "top 5" crowd based on prior performance, and the single best individual performer. In the experimental tasks, the selected "top 5" crowd performed the best in 21 tasks, compared to 14 for the whole crowd and 5 for the best individual. For economic forecasts, the selected "top 5" crowd out-performed the others in 34 cases, while the whole crowd was best in 15, and the top individual in only 1. These findings highlight that a small group of experts can outperform a large, mixed-ability group. They also show that crowds generally outperform the best individual expert.

The second study, by Mellers et al. (2014), examined predictions of real-world political events, such as: "Will Italy's Silvio Berlusconi resign, lose re-election/confidence vote, or otherwise vacate office before 1 January 2012?". Participants, all with at least a bachelor's degree, made predictions during two rounds: 2011–2012 and 2012–2013. The research-ers evaluated forecasts from individuals and teams, some of whom were trained in how to make good forecasts and others untrained. In Year 2, the top 2% of Year 1 performers were placed in elite "superforecaster" teams of 12. The elite teams significantly outperformed the other teams, who in turn outperformed the individuals.

Both studies demonstrate that small crowds selected for a high level of expertise can outperform larger, mixed-ability crowds and even top indi-vidual performers. Expertise within a crowd, rather than size alone, is important for better decision-making.

2.2   Cognitive Diversity

Cognitive diversity refers to the different ways people think and process information, including their knowledge, beliefs, skills, goals, and values (Sulik et al., 2022). This differs from sociodemographic diversity, which involves factors like age, gender, and ethnicity.

Research shows that cognitively diverse crowds make better decisions, especially when members' judgments are as independent (or negatively correlated) as possible while still being highly skilled (Davis-Stober et al., 2014; Page, 2007). For example, some tasks require multiple types of knowledge. A group where members complement each other's expertise will perform better than one where everyone shares the same knowledge. Conversely, if members' judgements are highly similar (positively corre-lated), their errors won't cancel out. In the extreme case, a crowd of "clones" performs no better than an individual.

Two studies are summarized here to illustrate the value of cognitive diversity. Keck and Tang (2020) tested cognitive process diversity in tasks like estimating historical dates, predicting soccer game outcomes, and guessing heights from photos. Participants were instructed to use different strategies, either an analytic approach or an intuitive one. It was found that crowds with a mix of strategies outperformed groups with only one approach.

Shi et al. (2019) studied the effect of political diversity in Wikipedia editors on the quality of articles. Quality was judged using Wikipedia's 6-category quality scale. Articles created by ideologically diverse groups (liberal and conservative editors) were found to be of higher quality than those made by homogeneous groups, especially when covering political and social issues. For political articles, having diverse groups of editors increased the odds of a higher-quality article by over 18 times. The reason for this was that disagreements within these groups led to more focused debates and robust edits.

Sociodemographic diversity has less of an impact on quality of judgements. De Oliveira and Nisbett (2018) explored diversity in age, gender, and ethnicity in decision-making tasks. They found minimal effects unless these factors were linked to cognitive diversity. Sociodemographic diversity alone rarely improved outcomes.

Cognitive diversity has the biggest impact on complex problems such as those found in science. Sulik et al. (2022) noted that science involves a range of subtasks, conflicting perspectives, and creativity, making it a prime example of where diversity improves outcomes.

In summary, cognitive diversity enhances the quality of group judgements, particularly in complex tasks, while sociodemographic diversity only helps when it aligns with cognitive differences.

2.3   Independence of Judgements

For the wisdom-of-crowds effect to work, individuals must make independent judgments—meaning their decisions aren't influenced by others. If judgments are not independent, errors can align in the same direction, creating systematic biases. Two common ways independence can be compromised are through groupthink and herding.

Groupthink happens when members of a small, close group accept what they believe is the group consensus, even if they have doubts. This term, introduced by Janis (1972), was originally linked to poor decisions in foreign policy, like the US-supported Bay of Pigs invasion of Cuba in 1961. Later research showed that groupthink is common in many decision-making scenarios, especially when members strongly identify with the group, feel social pressure to conform, or doubt their own decision-making ability (Baron, 2005).

Herding refers to copying others' behaviours, often seen when individuals make decisions one after another. For example, in financial markets, herding can cause bubbles or crashes. Herding can also occur with online judgements. For example, Muchnik et al. (2013) studied a social news site where users could vote on comments. When researchers gave comments an initial up-vote, it led to a positive-ratings "bubble", increasing final positive ratings by 25%. Initial negative votes, however, were usually neutralized by the crowd.

Another related factor that can compromise independence of judgements is the influence of a powerful individual, who other members may defer to even if this person's judgement is wrong. Groupthink, herding, and dominance by powerful individuals are less likely to occur if sharing of judgements is anonymous.

Several studies show how losing independence reduces the quality of group decisions (Locke & Anderson, 2015; Simoiu et al., 2019; Toyokawa et al., 2019). To give one example, in a study by Frey and Van de Rijt (2021), participants answered quiz questions either independently or after seeing other peoples' answers. When informed of the previous responses, group performance was worse, especially on harder questions.

2.4   Opportunity for Sharing

While independence is crucial for the wisdom-of-crowds effect, sharing information and discussion can sometimes enhance crowd decisions. For these benefits to occur, sharing must minimize biases like groupthink and herding. Studies showing improvements in judgements typically involve a process where individuals first make independent judgments, share information or discuss, and then make new independent judgments. These are aggregated for the final decision. This method differs from typical group decision-making, where individuals do not make initial independent judgments, identities are not anonymous, and powerful individuals may dominate.

One approach to sharing is to provide feedback about crowd judgements before a second round of independent decisions. Becker et al. (2017) studied this using an online "Intelligence Game". Participants estimated quantities (e.g. calories in food, candies in a jar) and were divided into groups with no feedback, decentralized feedback (equal influence

among participants), or centralized feedback (one individual had dispro-
portionate influence). Decentralized feedback improved judgement accuracy as individuals learned from the group. However, in centralized
networks, if the dominant person made errors, group estimates worsened.
Granovskiy et al. (2015) also found that feedback helped individuals adjust
outlier estimates closer to the correct value, improving group accuracy.

Discussion among group members has also been shown to enhance
judgements. Navajas et al. (2018) conducted an experiment with over
5,000 attendees at a live event. Participants initially provided individual
answers to general knowledge questions (e.g. what is the height of the
Eiffel Tower?). They then formed small groups, discussed, and provided
consensus answers for some of the questions. Later, they gave revised individual answers. The revised individual answers given after group discussions were more accurate than the initial ones. Similar benefits from
discussion have been observed in other studies (Dezecache et al., 2022;
Gürçay et al., 2015; Mellers et al., 2014; Mercier & Claidière, 2022).

3   Evaluation of Delphi Against the Conditions

If we generalize these findings from wisdom-of-crowds research to the
consensus methods reviewed in the previous chapter, validity is likely to be
greater under the following conditions:

1. The researcher selects a group of elite experts. The group should
   consist of top specialists in the field rather than a broad sample with
   only some expertise.
2. The group has cognitive diversity. Experts should have varied knowledge areas, disciplines, methodologies, and values.
3. Judgements are made independently. Methods like anonymous and
   independent voting should be used to minimize dominance and
   conformity pressures.
4. Sharing of expertise is encouraged. Opportunities are provided for
   experts to get feedback on other experts' judgements and discuss
   reasons for their judgements.

None of the methods summarized in Chap. 2 meets all four criteria.
However, the Delphi method and Nominal Group Technique come the
closest, as they involve both independence (Condition 3) and sharing
(Condition 4). In contrast, expert surveys lack sharing opportunities, and

methods like consensus conferences and expert working groups lack independent judgments. However, it is important in planning Delphi studies to also ensure that panellists have a high level of expertise (Condition 1) and that the panellists are diverse in their expertise (Condition 2).

It would be expected that consensus methods that better meet the conditions, such as the Delphi method, would have greater validity. However, there is only one study that has directly compared a range of the consensus methods. Graefe and Armstrong (2011) carried out a laboratory experiment comparing the Delphi method, the Nominal Group Technique, and prediction markets with traditional face-to-face meetings. In this experiment 227 students were assigned to small groups and asked to estimate the answers to ten factual questions (e.g. the population of Australia). Over all questions, the Delphi groups tended to do the best, but the differences between the methods were not statistically significant. When individual questions were examined, the Delphi groups did better than face-to-face meetings on two questions, with no difference on the other eight. Prediction markets were worse than face-to-face on three questions, while nominal groups and face-to-face were similar. We need to be cautious in drawing generalizations from this one experiment involving simple judgements, but it does give some support to the greater validity of Delphi groups compared to face-to-face meetings.

4   CONCLUSION

The wisdom-of-crowds literature gives support that the Delphi method can produce valid judgements. However, it also suggests areas where the design of Delphi studies can be improved, particularly in selecting experts for high levels of expertise and ensuring diversity of expertise. With these conclusions in mind, the next chapter walks the reader through the various steps in carrying out a Delphi study.

REFERENCES

Arechar, A. A., Allen, J., Berinsky, A. J., Cole, R., Epstein, Z., Garimella, K., Gully, A., Lu, J. G., Ross, R. M., Stagnaro, M. N., Zhang, Y., Pennycook, G., & Rand, D. G. (2023). Understanding and combatting misinformation across 16 countries on six continents. *Nature Human Behaviour, 7*(9), 1502–1513. https://doi.org/10.1038/s41562-023-01641-6

Baron, R. S. (2005). So right it's wrong: Groupthink and the ubiquitous nature of polarized group decision making. *Advances in Experimental Social Psychology, 37*, 219–253. https://doi.org/10.1016/S0065-2601(05)37004-3

Becker, J., Brackbill, D., & Centola, D. (2017). Network dynamics of social influence in the wisdom of crowds. *Proceedings of the National Academy of Sciences USA, 114*(26), E5070–E5076. https://doi.org/10.1073/pnas.1615978114

Brown, B., Cochran, S., & Dalkey, N. (1969). *The DELPHI method, II: Structure of experitments.* RAND.

Dalkey, N. (1969). *The Delphi method: An experimental study of group opinion.* RAND.

Dalkey, N., & Brown, B. (1971). *Comparison of group judgment techniques with short-range predictions and almanac questions.* RAND.

Davis-Stober, C. P., Budescu, D. V., Dana, J., & Broomell, S. B. (2014). When is a crowd wise? *Decision, 1*, 79–101. https://doi.org/10.1037/dec0000004

de Oliveira, S., & Nisbett, R. E. (2018). Demographically diverse crowds are typically not much wiser than homogeneous crowds. *Proceedings of the National Academy of Sciences USA, 115*(9), 2066–2071. https://doi.org/10.1073/pnas.1717632115

Dezecache, G., Dockendorff, M., Ferreiro, D. N., Deroy, O., & Bahrami, B. (2022). Democratic forecast: Small groups predict the future better than individuals and crowds. *Journal of Experimental Psychology: Applied, 28*(3), 525–537. https://doi.org/10.1037/xap0000424

Frey, V., & Van de Rijt, A. (2021). Social influence undermines the wisdom of crowds in sequential decision making. *Management Science, 67*, 4273–4286. https://doi.org/10.1287/mnsc.2020.3713

Galton, F. (1907). Vox populi. *Nature, 75*(1949), 450–451. https://doi.org/10.1038/075450a0

Graefe, A., & Armstrong, J. S. (2011). Comparing face-to-face meetings, nominal groups, Delphi and prediction markets on an estimation task. *International Journal of Forecasting, 27*(1), 183–195. https://doi.org/10.1016/j.ijforecast.2010.05.004

Granovskiy, B., Gold, J. M., Sumpter, D. J., & Goldstone, R. L. (2015). Integration of social information by human groups. *Topics in Cognitive Science, 7*(3), 469–493. https://doi.org/10.1111/tops.12150

Gürçay, B., Mellers, B. A., & Baron, J. (2015). The power of social influence on estimation accuracy. *Journal of Behavioral Decision Making, 28*, 250–261. https://doi.org/10.1002/bdm.1843

Janis, I. L. (1972). *Victims of groupthink.* Houghton Mifflin.

Jorm, A. (2025). *Expert consensus in science.* Palgrave Macmillan. https://doi.org/10.1007/978-981-97-9222-1

Keck, S., & Tang, W. (2020). Enhancing the wisdom of the crowd with cognitive-process diversity: The benefits of aggregating intuitive and analytical judgments. *Psychological Science*, *31*(10), 1272–1282. https://doi.org/10.1177/0956797620941840

Locke, C. C., & Anderson, C. (2015). The downside of looking like a leader: Power, nonverbal confidence, and participative decision-making. *Journal of Experimental Social Psychology*, *58*, 42–47. https://doi.org/10.1016/j.jesp.2014.12.004

Mannes, A. E., Soll, J. B., & Larrick, R. P. (2014). The wisdom of select crowds. *Journal of Personality and Social Psychology*, *107*, 276–299. https://doi.org/10.1037/a0036677

Martel, C., Allen, J., Pennycook, G., & Rand, D. G. (2024). Crowds can effectively identify misinformation at scale. *Perspectives in Psychological Science*, *19*, 477–488. https://doi.org/10.1177/17456916231190388

Mellers, B., Ungar, L., Baron, J., Ramos, J., Gurcay, B., Fincher, K., Scott, S. E., Moore, D., Atanasov, P., Swift, S. A., Murray, T., Stone, E., & Tetlock, P. E. (2014). Psychological strategies for winning a geopolitical forecasting tournament. *Psychological Science*, *25*(5), 1106–1115. https://doi.org/10.1177/0956797614524255

Mercier, H., & Claidière, N. (2022). Does discussion make crowds any wiser? *Cognition*, *222*, 104912. https://doi.org/10.1016/j.cognition.2021.104912

Muchnik, L., Aral, S., & Taylor, S. J. (2013). Social influence bias: A randomized experiment. *Science*, *341*, 647–651. https://doi.org/10.1126/science.1240466

Navajas, J., Niella, T., Garbulsky, G., Bahrami, B., & Sigman, M. (2018). Aggregated knowledge from a small number of debates outperforms the wisdom of large crowds. *Nature Human Behaviour*, *2*, 126–132. https://doi.org/10.1038/s41562-017-0273-4

O'Leary, D. E. (2017). Crowd performance in prediction of the World Cup 2014. *European Journal of Operational Research*, *260*, 715–724. https://doi.org/10.1016/j.ejor.2016.12.043

Page, S. E. (2007). *The difference: How the power of diversity creates better groups, firms, schools, and societies*. Princeton University Press.

Shi, F., Teplitskiy, M., Duede, E., & Evans, J. A. (2019). The wisdom of polarized crowds. *Nature Human Behaviour*, *3*(4), 329–336. https://doi.org/10.1038/s41562-019-0541-6

Simoiu, C., Sumanth, C., Mysore, A., & Goel, S. (2019). Studying the "wisdom of crowds" at scale. In *Seventh AAAI conference on human computation and crowdsourcing (HCOMP-19)*.

Sulik, J., Bahrami, B., & Deroy, O. (2022). The diversity gap: When diversity matters for knowledge. *Perspectives on Psychological Science, 17*(3), 752–767. https://doi.org/10.1177/17456916211006070

Surowiecki, J. (2004). *The wisdom of crowds: Why the many are smarter than the few.* Doubleday.

Toyokawa, W., Whalen, A., & Laland, K. N. (2019). Social learning strategies regulate the wisdom and madness of interactive crowds. *Nature Human Behaviour, 3*(2), 183–193. https://doi.org/10.1038/s41562-018-0518-x

CHAPTER 4

Steps in Carrying out a Delphi Study

Abstract This chapter deals with the practical details of how to design and implement a Delphi study. It covers setting an aim, selecting an expert panel, determining panel size, minimizing panel drop-out, constructing a questionnaire, providing additional evidence to panel members, piloting a questionnaire, providing feedback to the panel, defining consensus, deciding number of rounds, software for running Delphi studies, reporting Delphi studies, assessing the quality of Delphi studies and study pre-registration.

Keywords Delphi method • Aims • Expert panels • Panel size • Drop-out • Questionnaires • Piloting • Qualitative feedback • Quantitative feedback • Consensus • Survey rounds • Reporting studies • Preregistration

There is no single accepted way of carrying out a Delphi study. The Delphi method is best seen as a family of methods that shares some common features. When carrying out a Delphi study, a researcher has to make decisions about every step involved in designing and implementing the study. The aim of this chapter is to present the options available at each step and help the reader make informed choices.

© The Author(s) 2025
A. Jorm, *Using the Delphi Method to Establish Expert Consensus,*
Advancing Methods for Interdisciplinarity in the Social Sciences,
https://doi.org/10.1007/978-981-96-8357-4_4

1   SETTING AN AIM

The first step is to frame a clear aim which can be met by the Delphi method. The method can be used for a wide variety of consensus judgements. As described in Chap. 1, many uses fall into one of the following categories:

- Making judgements on facts where the evidence is imperfect
- Setting methodological standards
- Making predictions
- Defining foundational concepts
- Determining collective values
- Improving professional practice
- Improving policy

Below are examples of the aims of Delphi studies, drawn from the literature, which fit into one of the following categories:

- *Making judgements on facts where the evidence is imperfect*: "We sought consensus on dementia prevalence for all regions of the world, in 5-year age bands from 60 to 84 years, and for those aged 85 years and older" (Ferri et al., 2005, p. 2113).
- *Setting methodological standards*: "This study aimed to have international experts converge on a harmonized definition of whole hippocampus boundaries and segmentation procedures, to define standard operating procedures for magnetic resonance (MR)-based manual hippocampal segmentation" (Boccardi et al., 2015, p. 126).
- *Making predictions*: "This Delphi study aims to (1) generate a consensus among experts on a definition of food innovations and (2) forecast an overview of food innovations that are likely to be available to consumers in the next five years" (Zickafoose et al., 2022, p. 3).
- *Defining foundational concepts*: "The purpose of this consensus-based modified Delphi study was to achieve a more uniform terminology of fasting terms and regimens" (Koppold et al., 2024, p.18).
- *Determining collective values*: "This study aims to identify the top unanswered research priorities in the field of knee surgery using consensus-based methodology" (Ahmed & Metcalfe, 2024, p. 662).

- *Improving professional practice*: "A Delphi consensus study was conducted to define the knowledge, skills and attributes required of sonographers teaching point-of-care ultrasound to other health professionals in Australia and New Zealand" (Cormack et al., 2024, p. 1).
- *Improving policy*: "The aim of the Delphi process was to gain high agreement on the set of health states, events, and patient attributes to include in our [fall prevention intervention economic] model" (Saunders et al., 2023, p. 3).

1.1   Suggested Template for a Well-Framed Aim

Looking across the above examples, there are a number of elements that recur. These are as follows:

- Mention of consensus as the aim
- Specification of the issues that consensus is sought on
- Specification of the experts who will be involved
- Delphi specified as the method of achieving this

Putting these together, a template for a well-framed aim might be: "The aim is to assess the consensus of [list expert group/s] about [list issues] using the Delphi method". For example, "The aim is to assess the consensus of mental health professionals and people with lived experience of suicidality about suicide first aid strategies that members of the public can use to assist a suicidal person, using the Delphi method".

2   SELECTING AN EXPERT PANEL

Panellists need to be chosen for their expertise on a topic. According to the wisdom-of-crowds research summarized in Chap. 3, group judgements will tend to be better if the members are elite experts. A number of attributes can be used to indicate relevant scientific or professional expertise. I have previously reported that the following indicators are commonly used in consensus studies (Jorm, 2025):

- Professional qualifications, employment in the field and work experience
- Membership of a scientific or professional organization
- Peer-reviewed publications

- Invited participation in a specialized symposium
- Nomination by other experts

Niederberger and Spranger (2020) have summarized criteria that have been commonly used for the selection of experts in Delphi studies in the health sciences. These were organizational affiliation, recommendation by third parties, years of relevant experience, academic title, and number of publications. Many studies used multiple criteria.

In planning a Delphi study, it is best to have a clear definition of expertise relevant to the topic area and a sampling strategy for finding experts who meet this definition. Example 4.1 illustrates the use of selection criteria to ensure the required level of expertise.

Example 4.1 Criteria for Selecting Experts in a Delphi Study on Strategies That Parents Can Use to Prevent Adolescent Depression and Anxiety Problems

The authors of this study (Yap et al., 2014) described their selection criteria as follows: "In the current study, clinical and research experts in parenting and adolescent depression and anxiety formed the expert panel. … The panel was composed of experts with a minimum of five years experience in research investigating parenting and adolescent depression or anxiety (researchers) or clinical treatment involving parenting and adolescent depression or anxiety (clinicians). Researchers were identified based on corresponding authorship of articles included in our systematic literature review of parental factors associated with depression and anxiety disorders in adolescence. … Clinicians were identified via the Australian Psychological Society's *Find a Psychologist* website … by searching for Clinical Psychologists who listed their areas of expertise as 'Adolescents' and 'Anxiety & Phobias' or 'Depression'. Individuals known to the authors as having relevant clinical or research expertise were also invited to participate." (pp. 68–69)

According to the wisdom-of-crowds research summarized in Chap. 3, group judgements will tend to be better if there is cognitive diversity in the group. Such diversity often involves different types of disciplinary training or work experience, or types of methodological expertise. Schifano and Niederberger (2025) reviewed Delphi studies in the health sciences and found that most had diversity of expertise or professional background, with only 6% involving only one expert group. In terms of geographic origin, they found that 51% had national recruitment, 36% international, and 11% local or regional recruitment. However, whether or not recruitment should be international depends on the aims of the study; for aims that have a local focus (e.g. developing health services), an international panel might not be appropriate.

For Delphi studies that aim to establish consensus in areas where end-user values are important, such as services, policies, or resource allocation, expertise may come from "lived experience" as a patient, service user, consumer, or caregiver. A review of Delphi studies in the health sciences found that 27% of studies included affected populations such as patients as panel members (Schifano & Niederberger, 2025). Example 4.2 describes a study to develop consensus on what patients consider to be the most important outcomes of treatment.

Example 4.2 What Patients Consider to Be the Most Important Outcomes for Migraine Treatment

Outcome measures for studies on the effectiveness of migraine treatment are usually decided by clinicians. However, these may not be the most important to patients. To develop patient-centred outcome measures, Smelt et al. (2014) carried out a Delphi study with 300 migraine patients recruited from a large Dutch patient database. Participants were asked: "If a new medicine was developed for migraine attacks, what would you wish the effect of this medication to be?" The top priorities were found to be rapid relief of pain, preventing the migraine from becoming worse, preventing recurrence, restoring ability to function, and allowing the person to think clearly again within an hour. These outcomes differed from what was commonly measured in trials.

Some Delphi studies have multiple panels which may differ in their source of expertise or their values. Such studies can look for items where there is consensus across all the panels. Example 4.3 illustrates the use of consensus from multiple stakeholder groups, including informal caregivers, on a topic where the type of experience and values potentially play a major role.

Example 4.3 Use of Multiple Panels of Stakeholders to Identify Priorities for Caregiver Research in Cancer Care
Informal caregivers play an important role in assisting people with cancer. Lambert et al. (2019) carried out a Delphi study to identify priority topics for research on such caregivers. They recruited four different panels of stakeholders for this purpose: 103 cancer clinicians, 22 managers of relevant organizations (e.g. cancer care foundations), 63 researchers in psycho-oncology, and 61 caregivers of someone with cancer. They sought consensus across all four groups and found nine topics that achieved this. Examples of these include home care interventions and financial impact of burnout for caregivers and society.

Another type of expertise is cultural knowledge. This would be important when the consensus being established involves members of a particular cultural group or where the aim is to determine what cultural understanding is needed by providers of services to cultural minorities. Example 4.4 describes a Delphi study that used panellists who combined both professional and cultural expertise to develop guidelines for a cultural minority group.

Example 4.4 Development of Guidelines for Culturally Appropriate Responses to Mental Health Problems in Indigenous Australians
Hart et al. (2009) carried out a Delphi study to develop guidelines for providing mental health first aid to an Australian Aboriginal or Torres Strait Islander person who is experiencing a mental health crisis or developing a mental illness. Participants had to identify as an Aboriginal or Torres Strait Islander person, have experience working in the mental health field, and have excellent knowledge of Aboriginal mental health. There were 20–24 experts who participated in six independent Delphi studies to develop guidelines covering depression, psychosis, suicidal thoughts and behaviours, deliberate self-injury, trauma and loss, and cultural considerations. After completing the Delphi studies, participants were asked about the suitability of the Delphi method for Aboriginal people, with 83% agreeing that it can benefit Aboriginal people and recommending it for other projects for this group.

2.1   Some Pitfalls to Avoid When Recruiting Panel Members

To ensure cognitive diversity in a panel, panel members should not be selected in a way that favours experts with certain views or vested interests. Biased selection of this sort has been referred to as "panel stacking" by Kepp et al. (2024), who describe it as follows:

> Guided recruitment of similar views ("stacking") can occur when key members (eg, chairs or core groups) nominate panelists with strong views, preferences, or allegiances independent of evidence. Recruitment specifically because of expressed viewpoints and allegiance is a recognized major problem for guideline development. ... The issue can be exacerbated when stacked core group and panel members also choose the topics and phrasing of questions to be answered, weight the review or method toward their own knowledge rather than adhering to accepted evidence standards, and/or do not disclose conflicts of interest. (p. 4)

If the opinions or interests of the experts are a factor in recruitment, then this can result in a "manufactured consensus", which creates the appearance of a consensus where there is none. For example, Healy (2006) has cautioned about the manufacturing of consensus by pharmaceutical companies that are interested in promoting a new product:

> One of these methods involved the establishment of Delphi panels of experts. Delphi panels invite experts to consider clinical trial data and estimate the likely translation from the actually published randomized trial evidence to possible outcomes in clinical practice if the drugs are adopted widely. These outcomes are then costed by economists working for the manufacturing company. ... The invariable outcome of these proceedings has been sets of models indicating that treatment with newer agents costing 10 to 80 times more than older agents would in fact lead to savings. (p. 142)

Example 4.5 describes a Delphi study on COVID-19, which has been alleged to involve panel stacking.

Example 4.5 Potential Panel Stacking in a Delphi Study on the COVID-19 Threat to Public Health

Lazarus et al. (2022) carried out a Delphi study on how to end the threat of COVID-19 to public health. They recruited 386 experts covering multiple disciplines and organizations from 112 countries. The selection of the panelists involved a complex method of snowball sampling in which four co-chairs identified 40 experts who then proposed additional experts. The panel came to a consensus on 57 recommendations covering communication, health systems, vaccination, prevention, treatment and care, and inequities. Kepp et al. (2024) argued that the method of selecting panel members may have led to "panel stacking". To support this allegation, they searched for evidence that panel members had been involved in advocacy for COVID-19 elimination ("Zero-COVID"), which they argued was a minority position in 2022. Kepp et al. reported that at least 35% of the core panel members were Zero-COVID advocates and that this advocacy was often not declared. They concluded that: "In short, experts with strong, known preferences could select the topics, evidence, and final statements with little/no restraint from the community or impartial, systematic evidence synthesis" (p. 7). Whether or not the panel selection by Lazarus et al. (2022) did actually bias the consensus is unknown, but this example does illustrate a potential pitfall to be aware of when selecting panelists.

Another potential pitfall can occur with online recruitment of panellists. Online recruitment can be convenient for hard-to-access groups for which there is not a clear sampling frame, such as non-professional experts. This method of recruitment is fairly common. A review of recruitment of patient samples for Delphi studies found that 30% of studies recruited participants through a website or social media (Barrington et al., 2021). For some purposes, social media recruitment is arguably the most appropriate method, for example Delphi studies on best practice with online communication or online interventions for young people (Bailey et al., 2024; Biddle et al., 2022; Robinson et al., 2023). However, online recruitment methods can be severely affected by misrepresentation of eligibility (e.g. falsely claiming lived experience) or individuals creating multiple online identities in order to access financial incentives. In surveys using online recruitment, the percentage of responses that are fraudulent has been found to vary greatly, but in some cases has been greater than 80% of a sample (French et al.; Lei, 2024). Although these were not Delphi studies, the methods of recruitment are similar to those used in some Delphi studies. These experiences show the importance of verifying claims of expertise when online recruitment is used. A number of methods of verification and detection of fraud have been described in the literature (French et al.; Lei, 2024).

3   Determining Panel Size

There is comparatively little guidance on the appropriate size of a Delphi panel. In the Delphi literature, a wide range of panel sizes can be found. Niederberger and Spranger (2020) looked at the sample size of experts found in a number of different systematic reviews of Delphi studies in the health sciences. The number was found to vary from a low of 3 to a high of 731. One review found a median of 17 invited experts, and another found a mean of 40 experts in the first Delphi round. A review of more recent Delphi studies in the health sciences found a median of 31 experts in the first round of a Delphi study and 26 in the final round (Schifano & Niederberger, 2025).

There have been some studies estimating the stability of results according to Delphi panel size. A study by Akins et al. (2005) used a bootstrap method to examine the stability of responses of 23 experts in a Delphi study to create a patient safety tool. They found that their sample yielded reliable results and hypothesized that "Delphi surveys in other fields of study, conducted with a small number of experts with similar training and knowledge, would also yield reliable results" (pp. 10–11). However, whether these findings from a single study are generalizable to other Delphi studies is unknown.

A more recent study by Manyara et al. (2024) provided a sounder basis for choosing sample sizes. They investigated the replicability of Delphi consensus estimates using a much bigger dataset involving three Delphi studies with sample sizes of 175, 333, and 553, respectively. Manyara and colleagues used resampling with replacement to produce random subsamples of various sizes from each study. They then compared the median value of the survey ratings for each subsample with the medians for the full samples. For multistakeholder panels, a minimum sample size of 60–80 participants provided a high level of replicability in the results. For individual stakeholder groups (e.g. researchers, clinicians, or patients), a sample size of 20–30 was found to be sufficient.

While a larger panel will give a more precise estimate of consensus than a smaller one, sometimes there will be feasibility constraints on achieving a larger sample size. This may be the case if the research question is highly specialized and the number of experts on the topic is small. Although it might be possible to recruit a larger panel by relaxing the required level of expertise, the accuracy of group judgements may be reduced as a consequence.

There may also be feasibility constraints in running a Delphi study with a large panel. Delphi studies involve providing feedback to participants between survey rounds. For some studies, the feedback is quantitative (e.g. medians, per cent agreement), in which case providing feedback from a large sample is quite feasible. However, if the feedback is qualitative (e.g. rationales for ratings of each item), the cognitive load in reading and thinking about it will increase with sample size and this could become a disincentive to continued participation. To test whether this was the case, Belton et al. (2021) carried out an experiment in which they randomly assigned participants to Delphi panels varying in size from 7 to 19 members and for which opinion diversity was either low or high. Participants were asked 10 forecasting questions (e.g. "China and the US will resume trade talks designed to de-escalate the trade war" Yes/No) and to give rationales for their answers. They found that size of the panel and opinion diversity did not overload or disengage participants. However, the findings may have been different with a much larger panel (e.g. >100 panellists)

Another factor to consider is retention across survey rounds. The ideal is to have the full sample participate across all rounds, but drop-outs are to be expected. In planning a Delphi study, the initial sample size needs to be sufficient to allow for drop-outs so that reliable estimates can be made for subsequent survey rounds.

Putting all this together, it appears that some past Delphi studies have been too small to yield reliable estimates of consensus. Based on Manyara et al.'s (2024) study, 20–30 participants per stakeholder group will generally be sufficient, but it may be necessary to recruit more to allow for drop-out.

4   Minimizing Panel Drop-Out

Given that Delphi studies involve multiple survey rounds, panel drop-out across rounds can be a problem. This could potentially lead to bias in group judgements, for example, if panellists with dissenting views are more likely to drop out. A small number of studies have investigated panel drop-out and provided some guidance on how to minimize it.

To examine predictors of drop-out from Round 1 to Round 2, Gargon et al. (2019) analysed 31 Delphi studies to develop a core outcome set for clinical trials. The only predictors of drop-out they found were having a larger panel and more items to rate in Round 2. The authors thought that smaller panels might have better retention because they were more likely to be contacted personally by the research team rather than using more indirect methods. The effect of the number of items shows the importance of minimizing participant burden. Other features like the number of rounds, panel composition (single vs. multiple stakeholders), and international versus national recruitment did not predict drop-out, although it is possible that the number of studies was too small to detect other differences.

Boel et al. (2021) looked at the effect of two different invitation approaches for participants in subsequent rounds. They carried out a Delphi study in which panellists were randomly allocated to two groups. The first group received an invitation to participate in all rounds regardless of whether they had participated in the previous round. The second group only received an invitation if they had responded to the previous round. The response rate was higher for the "all rounds" group (61%) than the "responders only" group (46%). The researchers concluded that it was better to invite all panel members even if they had missed a round, as it resulted in a better representation of the views of the original panel.

Turnbull et al. (2018) did a survey of 70 panellists who participated in a five-round Delphi study with a retention rate of 90%, asking about their views and satisfaction with various aspects of the study. The panel was very broad, involving clinicians, researchers, former patients, and caregivers from many countries. The study made liberal use of email reminders, phone calls, and text messages to encourage continued participation. The post-study survey found that 96% of panellists were not bothered by these reminders, and 91% agreed that the time required to participate was appropriate. To achieve this level of engagement requires that a study be staffed appropriately.

The general conclusion from this limited research is that retention is likely to be better if there is a high level of engagement at recruitment and extensive efforts to follow up all panellists in subsequent rounds.

5   CONSTRUCTING A QUESTIONNAIRE

A questionnaire needs to be constructed for administration to the panellists. In some cases, the questionnaire will clearly flow from a very specific research aim. Example 4.6 illustrates this with a Delphi study to estimate the global prevalence of dementia.

Example 4.6 Estimates Required in a Delphi Study on the Global Prevalence of Dementia

To estimate the global prevalence of dementia, Ferri et al. (2005) provided 12 experts on the topic with a detailed document synthesizing the research evidence for each of 14 WHO regions, including a table that summarized the age-specific prevalence estimates for every available study. Experts were asked to review the evidence and make their own estimates for each of the regions for men and women combined in 5-year bands from 60–64 years up to 85+ years. The experts therefore had to provide 84 estimates of percent prevalence (6 age groups × 14 regions). As feedback, the experts were provided with a spreadsheet for every region giving individual estimates and the group response, which was the mean of the individual estimates. They were given the opportunity to revise their estimates after receiving feedback.

5.1   Methods for Generating Item Content

However, in most Delphi studies, the questionnaire content will not be specified so precisely by the research aim. Where this is the case, a method is needed for systematically generating the questionnaire items. It is important that the process of generating the questions is comprehensive and not biased by the interests of the researchers. If panellists are presented with a limited range of options or certain options are omitted, any consensus might be considered a manufactured one.

For complex research questions, there are several methods for generating the questionnaire content. A common one is to use qualitative research methods to elicit ideas from the Delphi panellists or other stakeholders. Example 4.7 describes a Delphi study to identify "Grand Challenges" in global eye health, which involved qualitative research with the panellists to generate content.

Example 4.7 Use of Qualitative Methods with Panellists to Generate Questionnaire Items in a Delphi Study on Grand Challenges in Global Eye Health

The aim of this study was to get consensus on the Grand Challenges in global eye health (Ramke et al., 2022). A Grand Challenge was defined as "a specific barrier, the removal of which would help solve an important health problem" (p. e33). To develop a list of challenges, panellists were given the open-ended question "What are the Grand Challenges in global eye health?" and invited them to propose up to five challenges and ways they could be addressed. Qualitative data analysis was used to group the responses into 21 categories which were organized into four broad themes: eye conditions, health systems, patient-related factors, and research. Steering group members reviewed the categories to check for duplications or ideas that had been missed from the original submissions. This resulted in a list of 85 unique challenges, from which panellists were asked to select and rank their top 20.

In some cases, the content is derived from qualitative interviews or focus groups with stakeholders who are not necessarily expert panellists. Example 4.8 illustrates this for a study to develop guidelines to assist adults to communicate with adolescents.

Example 4.8 Use of Focus Groups with Stakeholders to Generate Questionnaire Content for a Delphi Study on How Adults Can Communicate with Adolescents About Mental Health Problems

Fischer et al. (2013) carried out a Delphi study to develop guidelines for adults on how to communicate effectively with adolescents about mental health problems or other sensitive topics. To develop a questionnaire, they carried out a systematic literature search on tips to communicate with adolescents and two focus groups, one with past clients of a youth mental health service, and the other with clinicians and case managers from the service. In the past-client focus group, members were told: "Think about a time when you were a teenager when you were talking to an adult and you really felt that they were communicating effectively with you" (p. 3). The clinicians and case managers were told: "Reflect upon what you find works and does not work when communicating with young people. Think about how an adult can make a young person feel comfortable, heard and understood when talking to them, e.g. body language, what the adults can say and how they should say it" (p. 3). A researcher transcribed the audio recordings from each focus group and identified patterns of meaning to create potential themes. This pattern was repeated for the text found in the literature search. Data around each theme were used to create draft questionnaire items, which were discussed and refined by a working group to produce the final questionnaire. The questionnaire was administered to two panels: one of Youth Mental Health First Aid instructors and the other of young mental health consumer advocates.

The other commonly used method is to carry out a systematic literature search to find potential questionnaire content. Often this search will involve electronic databases of relevant scientific studies. This approach is illustrated in Example 4.9, which involved establishing a consensus on outcome measurements that should be used in pancreatic cancer trials.

Example 4.9 Use of a Systematic Literature Search to Develop a Delphi Questionnaire on Mandatory Measurements in Pancreatic Cancer Trials

Differences in reporting of potential confounding variables in cancer treatment trials make it difficult to compare findings between studies. To produce greater standardization in trials of pancreatic cancer, Ter Veer et al. (2018) carried out a Delphi study to establish a consensus on mandatory baseline and prognostic characteristics to be reported in future trials. To develop a potential list of characteristics to be measured, they carried out a systematic literature search for trials published between 2000 and 2016 using three electronic databases: MEDLINE, Embase, and the Cochrane Central Register of Controlled Trials. They also searched conference proceedings and clinical trial registries to ensure no trials had been missed. From the reports on these studies, the researchers extracted all reported baseline patient characteristics and all potential prognostic variables that had been included in multivariate analyses. In the first round of the Delphi study, 24 experts were presented with an overview of the results and asked to vote on which baseline and prognostic variables they wanted in the mandatory set.

If the output from a Delphi study is aimed at the general public or non-professionals, questionnaire items can be developed by systematically searching content from a general internet search engine or other lay sources rather than scientific databases. Example 4.10 illustrates this in a Delphi study on parenting strategies.

Example 4.10 Use of Google Search Engine to Develop Questionnaire Items for a Delphi Study on Parenting Strategies for Reducing the Risk of Childhood Depression and Anxiety Disorders
Yap and Jorm (2015) carried out a systematic review and meta-analysis of studies examining parental factors associated with childhood depression, anxiety, and internalizing problems. Although a number of reliable associations were found (e.g. with inter-parental conflict, lack of warmth, aversiveness), these did not provide specific guidance on what parents could do to reduce their child's risk. To fill this gap, Yap et al. (2015) carried out a Delphi study to establish a consensus on specific parenting strategies that reduce risk. To produce items for the Delphi questionnaire, they carried out systematic searches using Google search engines for several countries using the terms "(depression OR anxiety) AND prevent* AND (parenting OR parent*) AND (child OR childhood OR children)". The first 50 websites found were examined for any statements that could be useful to parents to prevent these mental health problems in their child. Relevant links to other sites were followed for additional content. A total of 723 recommendations were found and these were reduced to 471 unique ideas which were presented to a working group for review. The working group used these to produce 289 statements which were presented to panellists in round 1 of the Delphi study.

Even when a systematic search is used to generate questionnaire items, it is possible that some relevant content is missed. One method that has been used to fill any gaps is to ask panellists for any additional suggestions in the Round 1 survey. This approach was used in the study summarized above in Example 4.11. Any suggestions made were evaluated by a working group. If they involved a new idea or a better wording of an existing item, they were included for rating by panellists in Round 2.

Delphi studies vary greatly in the number of items that panellists must rate. Schifano and Niederberger (2025) found that Delphi studies in the health sciences varied between 1 and 289 items, with a median of 24.

Where the number of items is large, it can help the understanding of the rater to have the items grouped into thematic headings. This was the case for the study in Example 4.11, where the 289 items were organized under 22 headings (e.g. Warmth, Abuse and Neglect, Inter-parental Conflict).

Some studies randomize the order of questions, which may be appropriate when the task is to rank items for priority. However, it is rarely used in practice. Schifano and Niederberger (2025) reported that it is used in only 2% of recent Delphi studies in the health sciences.

5.2   Producing a Response Scale

When questionnaire items are administered to the expert panel, there needs to be a response scale. The simplest situation is with Delphi studies that ask for direct estimates of some quantity. Examples are:

- Age-specific prevalence (%) of dementia (Ferri et al., 2005)
- Commercial space demand (square footage per capita and number of stores per capita) in the Toronto region (Daniel & Hernandez, 2024)
- Dosing recommendations (mg/day) and estimates of clinically equivalent doses for 26 antipsychotic drug formulations (McAdam et al., 2023)
- Predicted number of visitors to an Expo to be held in Korea (Lee et al., 2008)
- Probability of species extinction in the wild in 20 years' time (Geyle et al., 2021)

Where the aim of the study is to produce a quantitative estimate, the researcher may also be interested in the uncertainty around that estimate. One of the Delphi variants, the IDEA Protocol, is specifically designed to make such estimates (Burgman, 2016; Hemming et al., 2018). It involves asking experts for upper and lower bounds, a best guess, and a rating of how sure they are. Example 4.11 illustrates how such estimates were collected in a natural resource management study.

> **Example 4.11 Eliciting Quantitative Judgements and Their Uncertainties**
>
> Hemming et al. (2018) used the IDEA Protocol to estimate 14 future abiotic and biotic events on the Great Barrier Reef. Questions were worded quite precisely to avoid any ambiguity. An example is: "How many tonnes of Coral Trout will be caught in Queensland by the Commercial Line Fishery in April 2016?". The questions on the estimates were as follows:
>
> - "Realistically, what do you think the lowest plausible catch (tonnes) of coral trout will be?"
> - "Realistically, what do you think the highest plausible catch (tonnes) of coral trout will be?"
> - "Realistically, what is your best guess for the catch (tonnes) of coral trout?"
> - "How confident are you that your interval, from lowest to highest, could capture the reported catch (tonnes) of coral trout? Please enter a number between 50 and 100%".
>
> The upper, lower, and confidence estimates were standardized using linear extrapolation to give an 80% credible interval for each expert.

However, most Delphi studies use Likert rating scales of agreement. A review of the use of the Delphi method in the health sciences found that these scales ranged from 2 to 11 points, with 5-point scales being the most common (40%), followed by 9-point scales (20%) (Schifano & Niederberger, 2025). The exact type of scale and the labels for the anchor points will vary depending on the subject matter and aims of the study. Table 4.1 gives some examples of scales used.

Although Likert scales involve multiple points, in practice, they are often dichotomized to analyse consensus. A simpler approach may therefore be to use a dichotomous response scale to start with. This is done in only a small minority of Delphi studies, but there are precedents, as shown in Table 4.2.

Other studies ask panellists to rank items, particularly where the Delphi study involves a prioritization. The disadvantage of using a Likert scale for

Table 4.1   Examples of Likert rating scales used in Delphi studies

Study	Aim	Judgement task	Rating scale used
Amaratunge et al. (2021)	Develop a checklist for reporting research using simulated patient methodology	Level of importance for inclusion in a simulated patient checklist	1 = Extremely important to 7 = Not at all important
Boccardi et al. (2015)	Develop a harmonized protocol for hippocampus segmentation on magnetic resonance	Level of agreement with including specific anatomical structures	1 = Minimum agreement to 9 = Maximum agreement
Cao et al. (2024)	Develop recommendations for clinical staging of patients newly diagnosed with breast cancer	Agreement with recommendation	Agree with statement as is, Agree with statement with edits, Do not agree with statement, Abstain
Chalmers et al. (2020)	Develop guidelines for mental health first aid after a potentially traumatic event	How important the item is for inclusion in guidelines	Essential, Important, Don't know/ depends, Unimportant, Should not be included
Cormack et al. (2024)	Define the competencies required by sonographers teaching ultrasound interprofessionally	Agreement with competency	Agree, Neutral, Disagree, Unable to comment
Liu et al. (2020)	Develop an extension to the CONSORT Statement for reporting clinical trials with an AI component	Importance of each reporting item	9-point scale: 1–3 not important, 4–6 important but not critical, 7–9 important and critical
Saunders et al. (2023)	Develop a fall prevention economic model	Impact of potential health states/events	0 = No impact to 5 = Very strong impact
Zickafoose et al. (2022)	Forecasting food innovations	Likelihood of consumer availability of a food innovation in next five years	1 = Strongly disagree to 5 = Strongly agree

assessing priorities is that panellists can give very high ratings to many items. With ranking, on the other hand, they are forced to choose among their top priorities. However, the disadvantage is that ranking is only feasible with a limited number of items. It may be possible to rank ten items, but the cognitive load of ranking hundreds of items is prohibitive. If the aim of the Delphi study is to agree on a small number of highest priorities,

Table 4.2 Examples of dichotomous scales used in Delphi studies

Study	Aim	Judgement task	Response scale used
Swedo et al. (2022)	Develop a consensus definition of misophonia	Agreement with a definitional statement	Agree/Disagree (also had an "Insufficient information" option)
ter Veer et al. (2018)	Develop a consensus statement on mandatory measurements in pancreatic cancer trials	Whether a baseline or prognostic variable should be included	Yes/No
Yu et al. (2023)	Develop an eye care competency framework	Whether competency statements should be kept or changed	Agree/Disagree

then the effort in ranking items further down the list may also be unnecessary. One way of overcoming this is to ask panellists to only rank their top priorities up to a certain number and then leave the rest. For example, in prioritizing Grand Challenges in global eye health (Example 4.7), Ramke et al. (2022) gave experts a list of 85 challenges to assess but asked them to only rank their top 10.

Some relevant data on rating versus ranking comes from an experimental study by Del Grande and Kaczorowski (2023) in which Delphi panellists were randomly assigned to use either rating or ranking to prioritize 16 items on the organization of primary cardiovascular care. The two methods showed high overall agreement in prioritization, particularly for the most favoured items. However, the ranking method was perceived as more difficult and tended to take longer. It is likely that the difference in difficulty and time would increase with a longer list of items to rank.

An alternative to ranking priorities is the value-weighting method. Panellists are given 100 points of hypothetical funding and asked to allocate these across areas. They could, for example, assign all 100 points to one priority or distribute the points across several. Example 4.12 illustrates the method.

Example 4.12 Use of the Value-Weighting Method to Allocate Priorities

Paul et al. (2011) carried out an expert consensus study to assess perceptions about psychosocial research priorities for adults with haematological cancers. They used a two-stage process. The first involved a modified Delphi study with researchers, health professionals, and patient representatives to achieve consensus on the range of research topics to be included in a questionnaire. The second stage involved administering this questionnaire to a sample of health professionals, patients, and carers. In this questionnaire, participants were asked to allocate 100 points of funding across various patient populations (e.g. newly diagnosed, in maintenance treatment, relapsed, receiving palliative care) and another 100 points across types of psychosocial research (e.g. identifying who is at risk of poor psychosocial health and who is resilient, testing the benefits of providing better information and education, evaluating the effectiveness of physical or psychological therapies). To identify the top priorities, the researchers looked for topics that received significantly higher points than the mean for all topics. For example, research on newly diagnosed patients was allocated a mean of 28.8 points compared to a value of 14.3 points if all options received equal allocation.

5.3   Collecting Qualitative Data Within the Questionnaire

Delphi questionnaires often collect qualitative comments from panellists in addition to quantitative ratings or rankings. This information can be used to provide feedback to panellists on rationales for judgements or to improve the item wording or item coverage in a subsequent survey round. This information is often collected through open text boxes, which can be associated with each item or with blocks of items. Example 4.13 describes how one study used text boxes to improve item wording and coverage.

Example 4.13 Use of Open Text Boxes in Delphi Questionnaire to Improve Item Wording and Coverage

Chalmers et al. (2020) carried out a Delphi study to develop mental health first aid guidelines for assisting a person after a potentially traumatic event. They described their use of open text boxes as follows:

> In Round 1, panel members were also asked to provide open-ended feedback after each section of the questionnaire. This allowed panellists to suggest helping actions that were not included in the first questionnaire. The authors reviewed this feedback and suggestions that contained original ideas were used to develop new helping statements to be included in the Round 2 questionnaire. Any statements that received feedback suggesting uncertainty in the interpretation of its meaning were re-phrased to make them unambiguous. These were included in the Round 2 questionnaire along with the statements from Round 1 that met the criteria to be re-rated. (p. 5)

In this study the open text box data was used by the researchers to improve or develop new items and not fed back directly to the panellists.

A less common use for qualitative comments collected as part of a Delphi questionnaire is to allow panellists to provide reasons for their judgements, which can be fed back to other panellists. Example 4.14 illustrates how this was done in one Delphi study using an online discussion board.

Example 4.14 Use of Open-Text Boxes to Record Reasons for Judgements

Radomski et al. (2022) carried out a Delphi study to develop a metric to detect low-value prescriptions in older adults using common forms of health care data. They used a software package (ExpertLens) to collect Likert scale ratings of items and free-text responses from panellists to explain their ratings in Round 1. Both the numeric ratings and free-text responses were fed back to the panellists in Round 2.

6   PROVIDING ADDITIONAL EVIDENCE
TO PANEL MEMBERS

Delphi panellists are recruited for their high level of relevant expertise. However, they may not necessarily have a detailed knowledge of the research relevant to the Delphi research question. For some Delphi questions there may be relevant research that uses experimental, longitudinal, cross-sectional, psychometric, or qualitative methods, which can inform expert judgements. It is good practice to carry out a systematic review of any evidence before doing a Delphi study and to provide the resulting review in a readily understandable form to the Delphi panellists to increase their expertise.

Where Delphi panellists are asked to provide direct estimates of some quantity, researchers would typically provide what data is available to inform those estimates. Example 4.7 above illustrates this for a Delphi study on the global prevalence of dementia, where panellists were provided with data from existing prevalence studies in different regions of the world.

Other studies use a systematic review or other narrative evidence summaries to inform Delphi panellists, as illustrated in Examples 4.15 and 4.16.

Example 4.15 Use of a Systematic Review to Inform Panellists in a Delphi Study on Intervention Priorities for Preventing or Reducing the Impact of Adverse Childhood Experiences
Sahle et al. (2022) carried out a Delphi study to establish expert consensus on priority interventions for preventing or reducing the impact of adverse childhood experiences in children under 8 years of age in Australia. To inform the panellists about relevant evidence, a systematic literature review was carried out. A total of 32 interventions were found to have been evaluated for their effectiveness. A short readable evidence summary was prepared. For each intervention, the summary included a description of the intervention, the population it was targeted at, the resources required for its implementation, the duration and intensity of the intervention, the level of supporting evidence, and any data on cost-effectiveness. Panellists were asked to rate each intervention for its priority for implementation in the Australian context.

Example 4.16 Use of "measure cards" to Inform Panellists in a Delphi Study on Core Outcome Measures for Clinical Research in Acute Respiratory Failure Survivors

Needham et al. (2017) carried out a Delphi study to get consensus on a minimum set of core outcome measures that should be included in clinical research studies that follow acute respiratory survivors after discharge. The researchers prepared a preliminary list of the 38 most-used measures and wrote standardized "measure cards" for each of these. The measure cards were written in a non-technical language and covered a range of attributes such as measurement properties, estimated time to complete, training requirements, licencing fees, and number of times used in previous research. Where available, the measure card included hyperlinks to additional information such as online examples or videos demonstrating the tests. Panellists were asked to rate each measure for importance on a 9-point scale and for suggestions for other potential measures not on the preliminary list.

A challenge in providing additional evidence is getting panellists to use it. In a follow-up survey of the Delphi panellists in the Needham et al. (2017) study (summarized in Example 4.16), Turnbull et al. (2018) asked them whether they had reviewed the "measure card" information provided. Only 27% reported reviewing all the measure cards and 30% reported reviewing only 0–50%. It is unclear whether this neglect of additional evidence was because of the burden of reading it or because the panellists already knew about the measures.

For some Delphi topics, there may be no relevant evidence. This is likely when the research questions involve ethical principles, cultural knowledge, or methodological standards. Delphi methods may also be deliberately used in situations where there is no evidence because of a lack of resources to carry out the research (Minas & Jorm, 2010). This may occur, for example, with healthcare planning in low-income countries and for cultural minorities within high-income countries. Although expert consensus is not a substitute for direct research on an issue, the consensus of a group is likely to provide a better basis for action than individual opinion.

7 PILOTING THE QUESTIONNAIRE

In addition to the items to be rated, a questionnaire will require instructions explaining the purpose of the study and the task requirements for participants, as well as any information required for informed consent. Once a complete questionnaire is drafted, it is important to pilot it to ensure comprehension and ease of use and to make changes accordingly.

Although Delphi panellists are asked to complete multiple questionnaires over a period of time, it should not be assumed that they will remember the instructions from a previous round. Turnbull et al. (2018) found that panellists often do not remember information presented over several months. These researchers emphasize the importance of repeating essential information in the stem of survey questions in each round.

Some Delphi studies also use cognitive interviewing to improve their questionnaire (e.g. Fan et al., 2024; Gibbins et al., 2017). Cognitive interviews involve a researcher asking participants to think aloud as they go through the questionnaire, which gives the researcher a respondent's perspective on their understanding of the questionnaire (Drennan, 2003). This is illustrated with Example 4.17.

Example 4.17 Use of Cognitive Interviewing When Piloting a Delphi Questionnaire

Gibbins et al. (2017) carried out a Delphi study to establish a consensus of experts in pharmacy and drug misuse on managing inappropriate use of non-prescription combination analgesics containing codeine. They described their use of cognitive interviewing when piloting the questionnaire as follows:

A pilot test, using a cognitive interview process, was conducted for every round to assess question readability and interpretation, and thereby improve the quality of the data. Between 2 and 4 local panellists (both pharmacist and non-pharmacist) were invited to participate in cognitive interviews. During the cognitive interviews the questionnaire was completed in front of a researcher, which allowed for comprehensive exploration of the participant experience of responding to the questions, and provided insight regarding patterns of interpretation of the questionnaire. After each cognitive interview the questionnaire was reviewed and amended, and then used in subsequent interviews. Once successive cognitive interviews indicated that no adjustments to the questionnaire were required, the questionnaire was finalized and released. (p. 371)

8   PROVIDING FEEDBACK TO THE PANEL

Chapter 3 concluded that an opportunity for sharing is one of the conditions that facilitate better group judgements. In Delphi studies, sharing occurs through feedback following the Round 1 questionnaire (and possibly following subsequent rounds), with the opportunity for panellists to then change their responses. The feedback provided may be quantitative or qualitative. Quantitative feedback involves statistical summaries of ratings from the panel. Qualitative feedback may involve sharing of reasons for judgements or using comments from the panel to create new items or changes to existing items. Delphi studies differ in the degree to which they emphasize these various types of feedback. Such differences in emphasis go back to the earliest Delphi studies at the RAND Corporation in the 1950s (Dayé, 2018).

According to a review of 48 Delphi studies to develop health care quality indicators, 58% gave quantitative feedback, 2% gave qualitative feedback, and 40% gave both quantitative and qualitative feedback (Boulkedid et al., 2011). Another review, covering Delphi studies in palliative care, reported on feedback provided in 18 studies, with 44% providing quantitative, 33% qualitative, and 22% both types of feedback (Jünger et al., 2017).

8.1   *Quantitative Feedback*

For quantitative feedback, the statistics provided can vary depending on the nature of the response scale. For scale ratings or estimates of quantities it is common to use some measure of central tendency (mean, median) and dispersion (minimum and maximum, interquartile range, standard deviation, frequency distribution). For categorical judgements or dichotomized responses from a Likert scale, percent endorsement is commonly used.

An assumption of the Delphi method is that providing feedback will persuade some panellists to change their responses and that such changes will result in better-quality group judgements. Barrios et al. (2021) analysed data from 628 panellists from five Delphi studies on mental health with the same design to see how much change occurred. After quantitative feedback they found that only 11% of judgements changed. When change did occur, it depended on the level of group endorsement that was fed back. If the group had at least 75% consensus about a judgement, then the dissenters were more likely to change towards the consensus on re-rating.

However, if there was less than 75% consensus, there was a tendency to move away from the majority view.

Some Delphi studies giving quantitative feedback present both the response of the group together with the panellist's individual response, whereas others give only the summary of the group. Boulkedid et al. (2011) reported that only 39% of the studies they reviewed gave individual feedback, which they argued was a limitation, as this is necessary for the individual panellist to compare themselves with the rest of the group. Some data on whether providing individual feedback makes a difference comes from an experiment by Meijering and Tobi (2018). They randomly assigned Delphi panellists to either receive summary statistics for the group and a summary of rationales given, or these summaries plus their own individual rating from the previous round. The panellists who received individual feedback were less likely to change their ratings following feedback and less likely to conform to the majority group response. However, whether this affected the quality of judgements is unknown.

Other research has examined the effects of feedback when the expert panel involves different groups of stakeholders. In a Delphi study with two stakeholder groups (health professionals and patients), Brookes et al. (2016) randomly assigned panellists to receive feedback from their peer stakeholder group only or from both stakeholder groups separately. Although the difference in feedback did not affect the percentage of items for which ratings were changed, consensus between professionals and patients was greater when they received feedback from both groups. Similarly, a study by MacLennan et al. (2018) randomly assigned professional and patient panellists to receive feedback from peers only, from both stakeholder groups separately, or both groups combined. They found no meaningful difference between the feedback groups but cautioned that this could be because of the high level of existing agreement at Round 1. A third study, by Campbell et al. (1999), did a similar experiment in a Delphi study where panellists were either family physicians or health care managers. They randomly assigned panellists to either receive feedback from their own professional group or from the whole panel. Managers were found to give higher ratings than physicians, but those receiving feedback from both groups were found to be influenced by the other group in their re-ratings. Taking these findings together, having stakeholder diversity in a panel does potentially have an influence on re-ratings.

8.2   Qualitative Feedback

Qualitative feedback is more complex to present than quantitative feedback. It could be in the form of a complete listing of comments made by panellists or a summary of the arguments and ideas. A major challenge in providing qualitative feedback is that the volume of comments can become very large and more difficult to process as the size of the panel and the number of items to be rated increases. If a summary is provided, this would require the use of qualitative research methods to produce a list of themes. In a review of Delphi studies in the health sciences, Schifano and Niederberger (2025) found that the most reported methods for analysis of qualitative data were content analysis (14%), thematic analysis (13%) and other methods (2%). It was concerning that most studies (62%) did not report or were unclear about the method used.

Where qualitative methods have been used, they have not been specific to the analysis of arguments presented by panellists in Delphi studies. To fill this gap, Niederberger and Homberg (2023) have proposed a method of analysing free-text responses from Delphi studies, called AQUA (Argument-based Qualitative Analysis strategy). However, its use has so far been limited to the developers' own work. The interested reader should consult their article for details on how to use the method.

Another use of qualitative feedback is to give ideas for additional Delphi items or changes in wording of existing items that can be put to the panel in a subsequent round. This is a way in which new ideas can be indirectly shared among the panel members.

8.3   Combined Quantitative and Qualitative Feedback

One review of Delphi methods has argued that panellists should be given both a statistical summary of the ratings and a summary of all comments received, as this will improve the quality of judgements (Boulkedid et al., 2011). Example 4.18 describes a Delphi study that used both types of feedback.

Example 4.18 A Delphi Study Providing Both Quantitative and Qualitative Feedback to Panellists

Niederberger et al. (2024) carried out a Delphi study to develop recommendations for standardized reporting of Delphi studies in the social and health sciences. The Delphi study involved three rounds. In Round 1, the expert panellists responded to a questionnaire which was developed based on a systematic review of the literature and discussions among a Delphi researcher network. The items of the questionnaire were organized under topics. The importance of reporting each item was rated in a 7-point Likert scale and panellists were provided with free-text boxes for comments at the end of each topic. In the Round 2 survey, panellists received quantitative feedback in the form of percentage of ratings of 6–7 on the scale, mean, and standard deviation. The panellists were also able to see their own responses from the previous round. Qualitative feedback was provided in the form of a summary of the arguments made in the text boxes. To produce this summary, arguments were extracted and categorized by topic using a specific qualitative analysis strategy—Argument-based Qualitative Analysis (AQUA). The Round 2 questionnaire included items that did not achieve consensus in Round 1 and items that were reworded based on comments from the previous round. For the Round 3 questionnaire, panellists again received quantitative feedback, but as there were so few new arguments in the free-text boxes, qualitative feedback was seen as unnecessary. The output from this study was the DELPHISTAR reporting checklist, which is described in more detail in Sect. 12 of this chapter.

Some research has been carried out on the effects of different types of feedback, but the findings have been inconsistent. Two studies looked at the accuracy of Delphi groups on short-term forecasting tasks in which they received either quantitative or qualitative feedback between rounds. One study found that qualitative feedback on reasons for judgements tended to increase group accuracy, but the difference from quantitative feedback was not statistically significant (Rowe & Wright, 1996). By contrast, a later study by the same researchers found that quantitative feedback improved group accuracy between rounds, but qualitative feedback did not (Rowe et al., 2005). A third study compared forecasting accuracy where the group received quantitative feedback only or both quantitative and qualitative feedback. This study found that adding qualitative feedback to reasons did not contribute to accuracy over quantitative feedback alone (Bolger et al., 2011).

Another experiment assigned panellists to receive either summary statistics plus rationales or rationales only (Meijering & Tobi, 2016), but in this study there was no external criterion of correctness to judge the effects of feedback on accuracy. Changes in ratings were found to be limited for both groups. However, the group receiving combined feedback was more likely to drop out, perhaps because of the greater burden involved. When panellists were asked what type of feedback they would prefer, both groups showed a strong preference for combined feedback.

Considering the available evidence, it is not possible to say that one type of feedback is better than the other. Researchers designing Delphi studies may need to consider other factors, such as feasibility and appropriateness to the judgement task, when deciding on the type of feedback.

8.4   Real-Time Feedback

A variant of the Delphi method, referred to as the "Real-time Delphi", gives immediate feedback without having defined survey rounds (Gordon & Pease, 2006). When a panellist provides a judgement, they are immediately presented with a summary of the group's responses to that point in time, which may be in the form of summary statistics, graphs, and comments. The panellist can also change their responses whenever they want.

The downside of this approach is that different panellists receive different feedback depending on how many others have responded ahead of them. For the early responders, the available feedback will be minimal, and they will have to log in later to receive more reliable feedback. It is also possible that any biases in feedback based on the initial responders will have an undue influence on later responders.

A comparison has been carried out between a Real-time Delphi study and a conventional Delphi study which were on the same topic (the future of the logistics industry) (Gnatzy et al., 2011). The researchers found that the final results did not differ according to the method used and there was no evidence of bias due to feedback in real time.

A more recent study by Quirke et al. (2023) randomly assigned panellists to a three-round Delphi or a Real-time Delphi method to establish consensus on a core outcome set for neonatal encephalopathy treatments.

They found that 28% of the outcomes were prioritized differently between the methods. Although fewer panellists changed their scores after feedback in the Real-time Delphi, the size of the difference was small. There was also some evidence that initial responders in the Real-time Delphi were less likely to change their ratings, but again the effect was small. An advantage of the Real-time Delphi is that it took less time to complete the study (14 vs. 20 weeks).

Real-time Delphi requires specialized software for implementation. A comparison of four available software tools has been published by Aengenheyster et al. (2017).

8.5 Incorporating Discussion as Part of Feedback

Some studies incorporate discussions among panellists as part of their feedback. Although this adds to the richness of information shared, it does potentially reduce the independence of judgements if contributions are not anonymous. Example 4.19 describes a study that used discussion in this way.

> **Example 4.19 A Delphi Study Involving Discussion Among Panellists Between Survey Rounds**
> Duijzer et al. (2024) carried out a Delphi study to reach a consensus on a clinical decision framework for the management of liver cyst infections. The expert panel were 24 medical specialists from 9 countries. The Delphi study had three rounds. Round 1 involved an online survey which had items based on a literature review. Round 2 involved an online panel meeting which was held to discuss items that did not meet the consensus (defined as 75% agreement) and newly suggested themes. Round 3 involved another online survey which was informed by the discussion in Round 2. The panellists were anonymous in Rounds 1 and 3, but identifiable to each other in the Round 2 discussion.

A specific Delphi variant that uses a discussion between two rounds is the RAND/UCLA Appropriateness Method (Charles et al., 2022; Fitch et al., 2001). The purpose of this method is to develop a consensus on the appropriateness of using a particular medical intervention when there is insufficient evidence from randomized controlled trials to guide this judgement. An intervention is defined as appropriate when the expected health benefit exceeds the expected negative consequences by a large enough margin to justify its use. After a review of the relevant evidence, clinical scenarios are developed, then the experts rate whether the intervention is appropriate for each scenario. In Round 1, the panellists make independent ratings without interaction. In Round 2, they meet for 1–2 days (often in person) to discuss the Round 1 results under the leadership of an experienced moderator. The discussion particularly focuses on areas of disagreement. Following this discussion, items are re-rated independently again.

9   DEFINING CONSENSUS

There is no standard for defining consensus in a Delphi study. The definition will vary depending on the aim of the study and the response scale used. If the judgement is a dichotomous one (e.g. agree/disagree) or a rating above a cutoff on a Likert scale (e.g. 6–7 on a 7-point scale), then per cent agreement would be appropriate. If the aim is to estimate some quantity (e.g. global prevalence of dementia, drug dose recommendation) or a score on a Likert scale, then a measure of central tendency (e.g. mean or median) or dispersion (e.g. interquartile range, and standard deviation) would be appropriate. The threshold for defining consensus on these response scales will also vary depending on the strength of consensus that the researcher is aiming for.

Reviews of Delphi studies covering a range of topics have reported that the most used definitions of consensus are per cent agreement, a measure of central tendency (particularly the median), or some combination of these (Boulkedid et al., 2011; Diamond et al., 2014; Foth et al., 2016; Jünger et al., 2017; Niederberger & Spranger, 2020; Schifano & Niederberger, 2025). Measures of dispersion are more rarely used.

Where per cent agreement has been used, the cutoffs for defining consensus have varied considerably from study to study. However, a review of the literature found cutoffs of 60% or higher in most cases (Niederberger & Spranger, 2020). One review reported that the median cutoff was 75% agreement (Foth et al., 2016), another review found that 75% and 80% agreement were the most used (Jünger et al., 2017), and a third review found that 70% was the most common, followed by 80% and 75% (Schifano & Niederberger, 2025). Some studies using per cent agreement involve multiple stakeholder groups, in which case the consensus threshold can be applied for each of these groups separately or for all groups combined. Example 4.20 illustrates this with a Delphi study that required a high percentage agreement from two stakeholder groups.

Example 4.20 Use of Per Cent Agreement from Multiple Stakeholder Groups to Define Consensus
Ross et al. (2014) carried out a Delphi study to develop guidelines on how family and friends can provide initial assistance to someone who is suicidal. The study involved two expert panels consisting of 41 suicide prevention professionals and 35 consumer advocates who had been suicidal in the past. The panellists rated 436 statements derived from a literature search and suggestions from panellists. The statements were rated on a Likert scale for importance to include in the guidelines. For a statement to be endorsed, at least 80% of both panels had to rate it as "Essential" or "Important" to include.

Some studies using a Likert scale specify a median to define consensus rather than per cent agreement over a cutoff. Example 4.21 illustrates this.

Example 4.21 Use of the Median on a Likert Scale to Define Consensus
Nilsson et al. (2023) carried out a Delphi study to develop the content and design for an internet-based education and support programme for patients waiting for a kidney transplantation. The experts were drawn from multiple stakeholder groups: 27 patients who had received a transplant, 6 significant others, and 34 healthcare professionals with either renal or transplant expertise. However, all stakeholder groups were treated as a single panel. Panellists were presented with a prototype of the intervention and had to rate the importance of content and design elements on a 5-point scale, with 1 indicating low importance and 5 very important. A median score of 3 or more was defined as consensus. All features had medians or 4 or 5, so were accepted as meeting the consensus.

In the above example, the researchers could have used a per cent agreement as an alternative that gives an equivalent result. Because the median is the 50th percentile, a median score of 3 or more is equivalent to at least 50% of the panellists giving a rating of 3 or more.

Use of measures of dispersion to define consensus is less common than for measures of central tendency but may be appropriate when the aim is to come to consensus on a particular score and a tight distribution of estimates around this score is required. This is illustrated in Example 4.22.

Example 4.22 Use of the Interquartile Range to Define Consensus
Sii et al. (2018) used the Delphi method to produce a consensus grading of glaucoma surgical complications. They asked 43 glaucoma experts to rate possible complications of various surgical procedures on a 1–10 scale, where 1 indicates no harm and 10 the worst possible surgical outcome. For each complication, the authors calculated the median and interquartile range. Consensus was defined as an interquartile range of ≤2 on the severity scale.

Another measure of dispersion that is occasionally used to define consensus in Delphi studies making quantitative estimates is the coefficient of variation (the ratio of the standard deviation to the mean). This coefficient is only meaningful for scales that have a true zero (ratio scales), which is rarely the case in Delphi studies. However, the interested reader can find an example in a study by Daniel and Hernandez (2024), which aimed to estimate the demand for retail space measured in square feet per capita.

Some Delphi studies have more complex definitions of consensus that involve multiple measures, as illustrated in Example 4.23.

Example 4.23 Use of Multiple Measures to Define Consensus
Thoomes et al. (2023) carried out a Delphi study to establish consensus on effective non-surgical treatment at different stages of cervical radiculopathy (pinched nerve). The experts were 27 professionals with a range of relevant backgrounds. They rated a list of treatment modalities proposed by a steering committee on a 5-point Likert scale ranging from 1 = Strongly Disagree to 5 = Strongly Agree. The consensus definition was different for each of the three survey rounds, but involved a combination of mean or median rating, per cent agreement, and interquartile range. At Round 3, the definition was ≥2 of the following: median rating of ≥4, interquartile range ≤1, and percentage agreement of ≥70%.

The disadvantage of using multiple measures is that the definition of consensus may lose any intuitive understanding for the reader. When the complexity is taken to an extreme, it can become almost incomprehensible, as demonstrated by the following quotation from a Delphi study on low-value prescribing in older people:

> [W]e looked for the presence of agreement by first calculating the interpercentile range (IPR) between the 70th and 30th percentiles of panelist scores. Next, we calculated the IPR adjusted for symmetry with the following equation: IPRAS = 2.35 + (AI × 1.5), where AI represented the asymmetry index, which is defined as the distance between the central point of the IPR and 5, the central point of the 9-point Likert scale used by panelists to rate each candidate low-value prescribing criteria. If the IPR was greater than the IPR adjusted for symmetry, then there was no agreement. If the IPR was less than the IPR adjusted for symmetry, then there was agreement. For the low-value prescribing criteria wherein agreement was found, we characterized their scientific validity and clinical usefulness according to the median scores. The final metric included only candidate low-value prescribing practices that showed panel agreement and received median scores of 6.5 or higher, which indicated both high scientific validity and high clinical usefulness. (Radomski et al., 2022, p. 5)

Whatever definition of consensus is used, this should be set in advance in the planning of the study and not done post-hoc during data analysis. Grant et al. (2018) have pointed out that it is easy to manipulate the percentage of items reaching consensus during analysis by changing definitions. Using data from eight Delphi studies, they were able to vary this percentage from 0% to 84% depending on the method of analysis.

The definition of consensus used has not always been given sufficient attention in the planning and reporting of Delphi studies. A review of 100 English-language Delphi studies by Diamond et al. (2014) found that although 98 of these aimed to assess consensus, only 72 provided a definition of consensus and 64 did this a priori. Similarly, in a review of 287 Delphi studies in the health sciences, Schifano and Niederberger (2025) found that only 36% reported that consensus was defined a priori, 3% that it was done a posteriori, and 61% were unclear or did not report on it. To remedy this situation, Diamond et al. (2014) proposed some key methodological criteria that should be reported in Delphi studies, including the definition of consensus and, if applicable, the threshold value that will be required for the Delphi to be stopped based on the achievement of consensus.

10   DECIDING NUMBER OF ROUNDS

There are two options for determining the number of rounds in a Delphi study. The first is to have a set number of rounds which is specified in the design of the study. The other is to continue the rounds until some predetermined consensus criterion is met. The former method is more common. Diamond et al. (2014) reviewed 100 Delphi studies and found that 71% had a set number of rounds, 24% continued until consensus was reached, 3% continued until there was stability of responses and the remaining 2% were unclear. In a review of Delphi studies in the health sciences, Schifano and Niederberger (2025) found that most (46%) were unclear or did not report the termination criterion, but if it was reported, the most common was the number of rounds (30%), followed by consensus reached (23%) and stability of judgements (9%).

The Diamond et al. (2014) review also found that 90% of Delphi studies had either two or three rounds, and this preponderance has been confirmed by other reviews (Humphrey-Murto et al., 2017; Jünger et al., 2017; Schifano & Niederberger, 2025). However, there is some inconsistency in the literature about what is counted as a round. Most Delphi studies have a questionnaire development stage, which can involve gathering ideas for items from panellists. This is then followed by a survey questionnaire that is administered to the panellists, followed by feedback and then administration of another questionnaire. Usually, the questionnaire administrations are referred to as Round 1, Round 2, and so on. However, some Delphi studies describe the initial idea-gathering stage as Round 1 (e.g. Ramke et al., 2022). A useful way to avoid this confusion is to adopt the terminology of Khodyakov et al. (2023) who use the term "Round 0" to refer to idea generation and "Round 1" to refer to the first questionnaire administration.

Example 4.24 illustrates the use of a fixed number of rounds, in this case three. The example has two notable features in the setting of rounds. Firstly, the study required the panellists to rate a large number of items. Having to re-rate all of these after feedback could be burdensome. Therefore, only items that came close to the consensus cutoff were re-rated, as changes in panel ratings after feedback are generally small. A second feature is that open-ended comments were only collected in Round 1, and these were used to generate new items that were included in Round 2. Only those new items that fell just below the consensus cutoff were then re-rated at Round 3.

Example 4.24 A Delphi Study with a Predetermined Number of Rounds

Berk et al. (2011) carried out a Delphi study to develop guidelines for caregivers of people with bipolar disorder. There were three expert panels: 45 caregivers, 47 people with bipolar disorder and 51 clinicians. At Round 1, the panellists were given a questionnaire with 626 items derived from a literature search, which were rated on a 5-point Likert scale ranging from "Essential" to "Should not be included". Panellists were also presented with evidence summaries to assist their judgements. To reach consensus, an item had to be rated as "Essential" or "Important" by at least 80% of all three panels. Open-ended questions allowed panellists to add comments and suggestions, which were used to generate additional items not included in the Round 1 questionnaire. In Round 2, panellists were presented with quantitative feedback showing the group percentages for each panel together with the panellist's own ratings for each item. They were asked to re-rate items that were close to consensus, defined as 70–79% endorsement from all three panels or by at least 80% endorsement by one of the panels. The Round 2 questionnaire also included the new items generated from comments in Round 1. Round 3 involved a much shorter questionnaire comprising the new items that met the re-rate criteria. As items were eligible to be re-rated only once, the study was terminated at Round 3.

Example 4.25 illustrates an alternative approach where the rounds are continued until a consensus criterion is met. This resulted in a longer study with five rounds.

Example 4.25 A Delphi Study That Continued with Rounds Until a Consensus Criterion Was Met

Boccardi et al. (2015) carried out a Delphi study to get a consensus on whole hippocampus boundaries and segmentation procedures when using magnetic resonance. There were 15 expert panellists who were presented with a questionnaire asking them to rate nine items about anatomical landmarks and give the reasons for their

(continued)

Example 4.25   (continued)

choices. Items were rated on a 9-point Likert scale, with scores of 6–9 classified as "Agree", 1–4 as "Disagree" and 5 as "Neutral". Voting rounds continued until there was a statistically significant difference in the number of panellists agreeing versus disagreeing. Most items achieved consensus at Round 2 or 3 but voting for one item continued to Round 4 and another to Round 5.

Humphrey-Murto and de Wit (2019) have cautioned about continuing rounds until consensus is met. They point out that panellists who have dissenting views may drop out or some panellists may agree just to make the rounds end, leading to a false consensus. Dayé (2018) has made a similar point, referring to the "fatigued expert":

[W]hat is the likely psychological effect of being repeatedly asked to revise one's answers in the face not of new evidence, but solely the opinion of the majority? Does such a procedure result in convergence? Or, rather, a mixture of annoyance and fatigue? (p. 861)

11   SOFTWARE FOR RUNNING DELPHI STUDIES

Contemporary Delphi studies are usually carried out over the internet. A review of Delphi studies in the health sciences carried out between 2018 and 2021 found that 80% were carried out online, while meetings between the panellists took place in only 16% (Schifano & Niederberger, 2025). In most studies, the surveys are administered with generic survey software. Schifano and Niederberger (2025) reported that Qualtrics was the most commonly used software for Delphi studies in the health sciences, making up 19% of studies, followed by SurveyMonkey (12%), REDCap (7%), Google Forms (5%), and LimeSurvey (4%).

There is also software that has been specifically developed for the purpose of running Delphi studies. Options include eDelphi (www.edelphi. org), SmartDelphi (www.smartdelphi.com), DelphiManager (www. comet-initiative.org/delphimanager), Surveylet (www.calibrum.com), Mesydel (www.mesydel.com) and RAND Expert Lens (www.rand.org/

pubs/tools/expertlens). Most of these applications require a fee, but some have basic or demo versions available for free. A table comparing the features of a range of software options can be found in Parks et al. (2018, p. 8).

12 Reporting Delphi Studies

A number of reviews have noted the inadequacies of reporting Delphi studies, with many studies reporting insufficient detail to allow an informed evaluation of their methodology (Humphrey-Murto et al., 2017; Jünger et al., 2017; Spranger et al., 2022). In response to this limitation, a range of guidelines have been developed specifying the elements that should be reported. A systematic review of reporting guidelines by Spranger et al. (2022) found ten guidelines, three of which have received limited use. Table 4.3 summarizes the remaining seven frequently used guidelines, plus two others that have been published since the Spranger et al. (2022) review. Most of these guidelines are specifically concerned with the reporting of Delphi studies, but two cover consensus studies more generally.

The two most recent guidelines (ACCORD and DELPHISTAR) involve a more thorough development process and include a greater number of checklist items. They are therefore likely to replace the earlier guidelines as reporting standards and are described in more detail here.

ACCORD (Accurate Consensus Reporting Document) was produced as a reporting guideline that could be used for consensus studies in biomedical research. To develop the guidelines, Gattrell et al. (2024) produced a preliminary checklist of 56 reporting items based on a systematic review of existing reporting guidelines and suggestions from a project steering committee. A Delphi study was carried out with a panel of 58 experts who reduced the preliminary checklist to a final list of 35 items. These items cover the article title (1 item), introduction (3 items), methods (21 items), results (5 items), discussion (2 items), and other information (1 item). More information about ACCORD and a copy of the checklist is available from https://www.ismpp.org/accord.

DELPHISTAR is an acronym for "Delphi studies in social and health sciences—recommendations for an interdisciplinary standardized reporting". It was developed from a systematic review and a Delphi study (Niederberger et al., 2024). The systematic review, covering the practice of Delphi studies and existing reporting guidelines, was used to identify an initial list of 65 items. The Delphi study involved 69 experts who reduced

Table 4.3   Guidelines for the reporting of Delphi studies or consensus studies more generally

Publication	Name of guideline	Focus	Method of development	Number of reporting elements
Hasson et al. (2000)	–	Delphi studies	Expertise of the three authors	17
Boulkedid et al. (2011)	–	Delphi studies for selecting healthcare quality indicators	Systematic review of reporting for 80 Delphi studies	19
Sinha et al. (2011)	–	Delphi studies to determine outcomes to measure in clinical trials	Systematic review of reporting for 15 Delphi studies	18
Diamond et al. (2014)	–	Reporting of consensus in Delphi studies	Systematic review of 100 Delphi studies	6
Waggoner et al. (2016)	–	Delphi, nominal group and consensus development panels	Literature review of 50 studies using one of the methods	4
Jünger et al. (2017)	CREDES	Delphi studies in palliative care	Systematic review of 30 Delphi studies	16
Humphrey-Murto et al. (2017)	–	Delphi, nominal group and other consensus methods in medical education research	Systematic review of 257 studies using one of the methods	13
Gattrell et al. (2024)	ACCORD	Consensus studies in biomedicine	Systematic review of existing literature on quality of reporting and Delphi study on checklist items	35
Niederberger et al. (2024)	DELPHISTAR	Delphi studies in social and health sciences	Systematic review of previous reporting guidelines and Delphi study on checklist items	38

these to a final list of 38 items. The checklist items cover title and abstract (3 items), context (7 items), method (20 items), results (4 items), and discussion (4 items). More information about DELPHISTAR and a copy of the checklist (in English and German) is available from https://www.equator-network.org/reporting-guidelines/delphi-studies-in-social-and-health-sciences-recommendations-for-an-interdisciplinary-standardized-reporting-delphistar-results-of-a-delphi-study/.

13   ASSESSING THE QUALITY OF DELPHI STUDIES

The checklists described above are aimed at improving the clarity of reporting but do not judge whether one methodological choice is better than another. However, there have been two proposals to assess published studies for quality. These may also be helpful in planning new studies.

Landeta and Lertxundi (2024) proposed quality indicators in three areas of a Delphi study: the quality of the expert panel (4 indicators), how information is obtained from the panellists (5 indicators) and the quality of interaction among the panellists (5 indicators). The indicators are rather general in their wording (e.g. "Configuration of a common context" and "Quantity and quality of information exchanged") and there are no specific criteria for scoring whether they have been met.

The second approach is the Delphi Critical Appraisal Tool (DCAT) proposed by Khodyakov et al. (2023). This consists of 16 items, 4 of which are described as "core" and the remaining 12 as "additional". The core items are as follows:

1. Did the research team appropriately employ *anonymity* in the Delphi study?
2. Did the research team appropriately employ *iteration* in the Delphi study?
3. Did the research team appropriately employ *statistical summaries of group responses* in the Delphi study?
4. Did the research team appropriately employ *controlled feedback* in the Delphi study? (Khodyakov et al., pp. 43–45).

These items are scored "Yes" if the assessor is confident that the standard is met, "No" if not and "Unsure" if there is insufficient information to make an assessment. The inter-scorer agreement on the DCAT is unknown. It is also not known whether studies with high scores on the DCAT produce better judgements. Nevertheless, the checklist may be helpful to work through when planning a Delphi study.

14   STUDY PRE-REGISTRATION

There has been a growing trend in the medical, social, and behavioural sciences to pre-register research studies. The growth in pre-registration has occurred because of concerns that negative findings or those contrary to the investigators' hypotheses may be withheld, leading to systematic bias in the literature (Lakens et al., 2024). There has also been concern that researchers may carry out many exploratory statistical tests and focus only on those that are statistically significant, reporting them as though they were hypothesized in advance (Nosek et al., 2018).

Similar concerns have also been expressed for Delphi studies. To support these concerns, Grant et al. (2018) analysed data from eight Delphi studies and found that the percentage of items that reached consensus varied from 0% to 84% depending on how consensus was defined in the analysis. These findings demonstrate that it is possible for researchers to analyse Delphi data in an exploratory way and potentially manipulate how many items are reported as reaching consensus. To prevent this from occurring, Grant et al. (2018) recommended that Delphi researchers pre-register their analysis plans.

Despite these concerns, most Delphi studies are currently not pre-registered. A possible reason is that most registries focus on randomized controlled trials. However, there are registries that take other research methods, including Delphi studies. These include OSF run by the Center for Open Science (https://help.osf.io/article/330-welcome-to-registrations) and the Research Registry run by the IGS Publishing Group (https://www.researchregistry.com/). Some journals also publish protocols for Delphi studies, as seen in examples by Lecours (2020) and Toulany et al. (2024).

15   CONCLUSION

It should be clear to the reader that there is no single Delphi method, rather a family of methods which involve the researcher in making informed decisions at each stage of the research. For some of the choices there is guidance from research or consensus of Delphi methodologists. Beyond such specific guidance, the general conditions for a valid consensus suggested by wisdom-of-crowds research provide a framework for making better methodological choices.

REFERENCES

Aengenheyster, S., Cuhls, K., Gerhold, L., Heiskanen-Schüttler, M., Huck, J., & Muszynska, M. (2017). Real-time Delphi in practice—A comparative analysis of existing software-based tools. *Technological Forecasting and Social Change, 118*, 15–27. https://doi.org/10.1016/j.techfore.2017.01.023

Ahmed, I., & Metcalfe, A. (2024). Research priorities of members of the British Association for Surgery of the Knee. *Bone and Joint Journal, 106 B*(7), 662–668. https://doi.org/10.1302/0301-620X.106B7.BJJ-2023-0691.R1

Akins, R. B., Tolson, H., & Cole, B. R. (2005). Stability of response characteristics of a Delphi panel: Application of bootstrap data expansion. *BMC Medical Research Methodology, 5*, 37. https://doi.org/10.1186/1471-2288-5-37

Amaratunge, S., Harrison, M., Clifford, R., Seubert, L., Page, A., & Bond, C. (2021). Developing a checklist for reporting research using simulated patient methodology (CRiSP): A consensus study. *International Journal of Pharmacy Practice, 29*(3), 218–227. https://doi.org/10.1093/ijpp/riaa002

Bailey, E., Bellairs-Walsh, I., Reavley, N., Gooding, P., Hetrick, S., Rice, S., Boland, A., & Robinson, J. (2024). Best practice for integrating digital interventions into clinical care for young people at risk of suicide: A Delphi study. *BMC Psychiatry, 24*(1), 71. https://doi.org/10.1186/s12888-023-05448-7

Barrington, H., Young, B., & Williamson, P. R. (2021). Patient participation in Delphi surveys to develop core outcome sets: Systematic review. *BMJ Open, 11*(9), e051066. https://doi.org/10.1136/bmjopen-2021-051066

Barrios, M., Guilera, G., Nuño, L., & Gómez-Benito, J. (2021). Consensus in the delphi method: What makes a decision change? *Technological Forecasting and Social Change, 163,* 120484. https://doi.org/10.1016/j.techfore. 2020.120484

Belton, I., Wright, G., Sissons, A., Bolger, F., Crawford, M. M., Hamlin, I., Taylor Browne Lūka, C., & Vasilichi, A. (2021). Delphi with feedback of rationales: How large can a Delphi group be such that participants are not overloaded, de-motivated, or disengaged? *Technological Forecasting and Social Change, 170,* 120897. https://doi.org/10.1016/j.techfore.2021.120897

Berk, L., Jorm, A. F., Kelly, C. M., Dodd, S., & Berk, M. (2011). Development of guidelines for caregivers of people with bipolar disorder: A Delphi expert consensus study. *Bipolar Disorders, 13*(5–6), 556–570. https://doi.org/10.1111/j.1399-5618.2011.00942.x

Biddle, L., Rifkin-Zybutz, R., Derges, J., Turner, N., Bould, H., Sedgewick, F., Gooberman-Hill, R., Moran, P., & Linton, M. J. (2022). Developing good practice indicators to assist mental health practitioners to converse with young people about their online activities and impact on mental health: A two-panel mixed-methods Delphi study. *BMC Psychiatry, 22*(1), 485. https://doi.org/10.1186/s12888-022-04093-w

Boccardi, M., Bocchetta, M., Apostolova, L. G., Barnes, J., Bartzokis, G., Corbetta, G., DeCarli, C., deToledo-Morrell, L., Firbank, M., Ganzola, R., Gerritsen, L., Henneman, W., Killiany, R. J., Malykhin, N., Pasqualetti, P., Pruessner, J. C., Redolfi, A., Robitaille, N., Soininen, H., et al. (2015). Delphi definition of the EADC-ADNI harmonized protocol for hippocampal segmentation on magnetic resonance. *Alzheimer's & Dementia, 11*(2), 126–138. https://doi.org/10.1016/j.jalz.2014.02.009

Boel, A., Navarro-Compán, V., Landewé, R., & van der Heijde, D. (2021). Two different invitation approaches for consecutive rounds of a Delphi survey led to comparable final outcome. *Journal of Clinical Epidemiology, 129*, 31–39. https://doi.org/10.1016/j.jclinepi.2020.09.034

Bolger, F., Stranieri, A., Wright, G., & Yearwood, J. (2011). Does the Delphi process lead to increased accuracy in group-based judgmental forecasts or does it simply induce consensus amongst judgmental forecasters? *Technological Forecasting and Social Change, 78*(9), 1671–1680. https://doi.org/10.1016/j.techfore.2011.06.002

Boulkedid, R., Abdoul, H., Loustau, M., Sibony, O., & Alberti, C. (2011). Using and reporting the Delphi method for selecting healthcare quality indicators: A systematic review. *PLoS One, 6*(6), e20476. https://doi.org/10.1371/journal.pone.0020476

Brookes, S. T., Macefield, R. C., Williamson, P. R., McNair, A. G., Potter, S., Blencowe, N. S., Strong, S., & Blazeby, J. M. (2016). Three nested randomized controlled trials of peer-only or multiple stakeholder group feedback within Delphi surveys during core outcome and information set development. *Trials, 17*(1), 409. https://doi.org/10.1186/s13063-016-1479-x

Burgman, M. A. (2016). *Trusting judgements: How to get the best out of experts.* Cambridge University Press.

Campbell, S. M., Hann, M., Roland, M. O., Quayle, J. A., & Shekelle, P. G. (1999). The effect of panel membership and feedback on ratings in a two-round Delphi survey: Results of a randomized controlled trial. *Medical Care, 37*(9), 964–968. https://doi.org/10.1097/00005650-199909000-00012

Cao, J. Q., Surgeoner, B., Manna, M., Boileau, J. F., Gelmon, K. A., Brackstone, M., Brezden-Masley, C., Jerzak, K. J., Prakash, I., Sehdev, S., Wong, S. M., Bouganim, N., Cescon, D. W., Chia, S., Dayes, I. S., Joy, A. A., & Henning, J. W. (2024). Guidance for Canadian breast cancer practice: National consensus recommendations for clinical staging of patients newly diagnosed with breast cancer. *Current Oncology*, *31*(11), 7226–7243. https://doi.org/10.3390/curroncol31110533

Chalmers, K. J., Jorm, A. F., Kelly, C. M., Reavley, N. J., Bond, K. S., Cottrill, F. A., & Wright, J. (2020). Offering mental health first aid to a person after a potentially traumatic event: A Delphi study to redevelop the 2008 guidelines. *BMC Psychology*, *8*(1), 105. https://doi.org/10.1186/s40359-020-00473-7

Charles, K. R., Hall, L., Ullman, A. J., & Schults, J. A. (2022). Methodology minute: Utilizing the RAND/UCLA appropriateness method to develop guidelines for infection prevention. *American Journal of Infection Control*, *50*(3), 345–348. https://doi.org/10.1016/j.ajic.2021.12.012

Cormack, C. J., Childs, J., & Kent, F. (2024). Competencies required by sonographers teaching ultrasound interprofessionally: A Delphi consensus study. *BMC Medical Education*, *24*(1), Article 970. https://doi.org/10.1186/s12909-024-05933-x

Daniel, C., & Hernandez, T. (2024). What retail apocalypse? A Delphi forecast of commercial space demand in the Toronto region. *Journal of Retailing and Consumer Services*, *77*, 103670. https://doi.org/10.1016/j.jretconser.2023.103670

Dayé, C. (2018). How to train your oracle: The Delphi method and its turbulent youth in operations research and the policy sciences. *Social Studies of Science*, *48*(6), 846–868. https://doi.org/10.1177/0306312718798497

Del Grande, C., & Kaczorowski, J. (2023). Rating versus ranking in a Delphi survey: A randomized controlled trial. *Trials*, *24*(1), 543. https://doi.org/10.1186/s13063-023-07442-6

Diamond, I. R., Grant, R. C., Feldman, B. M., Pencharz, P. B., Ling, S. C., Moore, A. M., & Wales, P. W. (2014). Defining consensus: A systematic review recommends methodologic criteria for reporting of Delphi studies. *Journal of Clinical Epidemiology*, *67*(4), 401–409. https://doi.org/10.1016/j.jclinepi.2013.12.002

Drennan, J. (2003). Cognitive interviewing: Verbal data in the design and pretesting of questionnaires. *Journal of Advanced Nursing*, *42*(1), 57–63. https://doi.org/10.1046/j.1365-2648.2003.02579.x

Duijzer, R., Bernts, L. H. P., Geerts, A., van Hoek, B., Coenraad, M. J., Rovers, C., Alvaro, D., Kuijper, E. J., Nevens, F., Halbritter, J., Colmenero, J., Kupcinskas, J., Salih, M., Hogan, M. C., Ronot, M., Vilgrain, V., Hanemaaijer, N. M., Kamath, P. S., Strnad, P., et al. (2024). Clinical management of liver cyst infections: An international, modified Delphi-based clinical decision framework. *The Lancet Gastroenterology & Hepatology*, *9*(9), 884–894. https://doi.org/10.1016/S2468-1253(24)00094-3

Fan, T., Zhu, S., Wang, H., Dong, Y., Zhou, Y., Song, Y., Pan, S., Wu, Q., Smith, G. D., Li, Y., & Han, Y. (2024). Development and validation of the self-report symptom inventory of immune-related adverse events in patients with lung cancer. *Asia-Pacific Journal of Oncology Nursing, 11*(12), 100603. https://doi. org/10.1016/j.apjon.2024.100603

Ferri, C. P., Prince, M., Brayne, C., Brodaty, H., Fratiglioni, L., Ganguli, M., Hall, K., Hasegawa, K., Hendrie, H., Huang, Y., Jorm, A., Mathers, C., Menezes, P. R., Rimmer, E., & Scazufca, M. (2005). Global prevalence of dementia: A Delphi consensus study. *Lancet, 366*(9503), 2112–2117. https://doi. org/10.1016/s0140-6736(05)67889-0

Fischer, J. A., Kelly, C. M., Kitchener, B. A., & Jorm, A. F. (2013). Development of guidelines for adults on how to communicate with adolescents about mental health problems and other sensitive topics: A Delphi study. *SAGE Open, 3*(4), 2158244013516769. https://doi.org/10.1177/2158244013516769

Fitch, K., Bernstein, S. J., Aguilar, M. D., Burnand, B., LaCalle, J. R., & Lazaro, P. (2001). *The RAND/UCLA appropriateness method user's manual.* RAND. https://search.library.wisc.edu/catalog/999911619702121

Foth, T., Efstathiou, N., Vanderspank-Wright, B., Ufholz, L. A., Dütthorn, N., Zimansky, M., & Humphrey-Murto, S. (2016). The use of Delphi and Nominal Group Technique in nursing education: A review. *International Journal of Nursing Studies, 60*, 112–120. https://doi.org/10.1016/j. ijnurstu.2016.04.015

Gargon, E., Crew, R., Burnside, G., & Williamson, P. R. (2019). Higher number of items associated with significantly lower response rates in COS Delphi surveys. *Journal of Clinical Epidemiology, 108*, 110–120. https://doi. org/10.1016/j.jclinepi.2018.12.010

Gattrell, W. T., Logullo, P., van Zuuren, E. J., Price, A., Hughes, E. L., Blazey, P., Winchester, C. C., Tovey, D., Goldman, K., Hungin, A. P., & Harrison, N. (2024). ACCORD (ACcurate COnsensus Reporting Document): A reporting guideline for consensus methods in biomedicine developed via a modified Delphi. *PLoS Medicine, 21*(1), e1004326. https://doi.org/10.1371/journal. pmed.1004326

Geyle, H. M., Tingley, R., Amey, A. P., Cogger, H., Couper, P. J., Cowan, M., Craig, M. D., Doughty, P., Driscoll, D. A., Ellis, R. J., Emery, J. P., Fenner, A., Gardner, M. G., Garnett, S. T., Gillespie, G. R., Greenlees, M. J., Hoskin, C. J., Keogh, J. S., Lloyd, R., et al. (2021). Reptiles on the brink: Identifying the Australian terrestrial snake and lizard species most at risk of extinction. *Pacific Conservation Biology, 27*(1), 3–12. https://doi.org/10.1071/PC20033

Gibbins, A. K., Wood, P. J., & Spark, M. J. (2017). Managing inappropriate use of non-prescription combination analgesics containing codeine: A modified Delphi study. *Research in Social & Administrative Pharmacy, 13*(2), 369–377. https://doi.org/10.1016/j.sapharm.2016.02.015

Gnatzy, T., Warth, J., von der Gracht, H. A., & Darkow, I. (2011). Validating an innovative real-time Delphi approach – A methodological comparison between real-time and conventional Delphi studies. *Technological Forecasting and Social Change, 78*(9), 1681–1694. https://doi.org/10.1016/j. techfore.2011.04.006

Gordon, T., & Pease, A. (2006). RT Delphi: An efficient, "round-less" almost real time Delphi method. *Technological Forecasting and Social Change, 73*(4), 321–333. https://doi.org/10.1016/j.techfore.2005.09.005

Grant, S., Booth, M., & Khodyakov, D. (2018). Lack of preregistered analysis plans allows unacceptable data mining for and selective reporting of consensus in Delphi studies. *Journal of Clinical Epidemiology, 99*, 96–105. https://doi. org/10.1016/j.jclinepi.2018.03.007

Hart, L. M., Jorm, A. F., Kanowski, L. G., Kelly, C. M., & Langlands, R. L. (2009). Mental health first aid for Indigenous Australians: Using Delphi consensus studies to develop guidelines for culturally appropriate responses to mental health problems. *BMC Psychiatry, 9*, 47. https://doi.org/10.118 6/1471-244x-9-47

Hasson, F., Keeney, S., & McKenna, H. (2000). Research guidelines for the Delphi survey technique. *Journal of Advanced Nursing, 32*(4), 1008–1015.

Healy, D. (2006). Manufacturing consensus. *Culture, Medicine and Psychiatry, 30*(2), 135–156. https://doi.org/10.1007/s11013-006-9013-3

Hemming, V., Burgman, M. A., Hanea, A. M., McBride, M. F., & Wintle, B. C. (2018). A practical guide to structured expert elicitation using the IDEA protocol. *Methods in Ecology and Evolution, 9*(1), 169–180. https://doi.org/1 0.1111/2041-210X.12857

Hemming, V., Walshe, T. V., Hanea, A. M., Fidler, F., & Burgman, M. A. (2018). Eliciting improved quantitative judgements using the IDEA protocol: A case study in natural resource management. *PLoS One, 13*(6), e0198468. https:// doi.org/10.1371/journal.pone.0198468

Humphrey-Murto, S., & de Wit, M. (2019). The Delphi method-more research please. *Journal of Clinical Epidemiology, 106*, 136–139. https://doi. org/10.1016/j.jclinepi.2018.10.011

Humphrey-Murto, S., Varpio, L., Wood, T. J., Gonsalves, C., Ufholz, L. A., Mascioli, K., Wang, C., & Foth, T. (2017). The use of the Delphi and other consensus goup methods in medical education research: A review. *Academic Medicine, 92*(10), 1491–1498. https://doi.org/10.1097/acm.0000000000001812

Jorm, A. (2025). *Expert consensus in science.* Palgrave Macmillan. https://doi. org/10.1007/978-981-97-9222-1

Jünger, S., Payne, S. A., Brine, J., Radbruch, L., & Brearley, S. G. (2017). Guidance on Conducting and REporting DElphi Studies (CREDES) in palliative care: Recommendations based on a methodological systematic review. *Palliative Medicine, 31*(8), 684–706. https://doi.org/10.1177/0269216317690685

Kepp, K. P., Aavitsland, P., Ballin, M., Balloux, F., Baral, S., Bardosh, K., Bauchner, H., Bendavid, E., Bhopal, R., Blumstein, D. T., Boffetta, P., Bourgeois, F., Brufsky, A., Collignon, P. J., Cripps, S., Cristea, I. A., Curtis, N., Djulbegovic,

B., Faude, O., et al. (2024). Panel stacking is a threat to consensus statement validity. *Journal of Clinical Epidemiology, 173*, Article 111428. https://doi.org/10.1016/j.jclinepi.2024.111428

Khodyakov, D., Grant, S., Kroger, J., & Bauman, M. (2023). *RAND methodological guidance for conducting and critically appraising Delphi panels*. RAND Corporation.

Koppold, D. A., Breinlinger, C., Hanslian, E., Kessler, C., Cramer, H., Khokhar, A. R., Peterson, C. M., Tinsley, G., Vernieri, C., Bloomer, R. J., Boschmann, M., Bragazzi, N. L., Brandhorst, S., Gabel, K., Goldhamer, A. C., Grajower, M. M., Harvie, M., Heilbronn, L., Horne, B. D., et al. (2024). International consensus on fasting terminology. *Cell Metabolism, 36*(8), 1779–1794.e1774. https://doi.org/10.1016/j.cmet.2024.06.013

Lakens, D., Mesquida, C., Rasti, S., & Ditroilo, M. (2024). The benefits of pre-registration and Registered Reports. *Evidence-Based Toxicology, 2*(1), 2376046. https://doi.org/10.1080/2833373X.2024.2376046

Lambert, S. D., Ould Brahim, L., Morrison, M., Girgis, A., Yaffe, M., Belzile, E., Clayberg, K., Robinson, J., Thorne, S., Bottorff, J. L., Duggleby, W., Campbell-Enns, H., Kim, Y., & Loiselle, C. G. (2019). Priorities for caregiver research in cancer care: An international Delphi survey of caregivers, clinicians, managers, and researchers. *Supportive Care in Cancer, 27*(3), 805–817. https://doi.org/10.1007/s00520-018-4314-y

Landeta, J., & Lertxundi, A. (2024). Quality indicators for Delphi studies. *FUTURES & FORESIGHT SCIENCE, 6*(1), e172. https://doi.org/10.1002/ffo2.172

Lazarus, J. V., Romero, D., Kopka, C. J., Karim, S. A., Abu-Raddad, L. J., Almeida, G., Baptista-Leite, R., Barocas, J. A., Barreto, M. L., Bar-Yam, Y., Bassat, Q., Batista, C., Bazilian, M., Chiou, S. T., Del Rio, C., Dore, G. J., Gao, G. F., Gostin, L. O., Hellard, M., et al. (2022). A multinational Delphi consensus to end the COVID-19 public health threat. *Nature, 611*(7935), 332–345. https://doi.org/10.1038/s41586-022-05398-2

Lecours, A. (2020). Scientific, professional and experiential validation of the model of preventive behaviours at work: Protocol of a modified Delphi Study. *BMJ Open, 10*(9), e035606. https://doi.org/10.1136/bmjopen-2019-035606

Lee, C., Song, H., & Mjelde, J. W. (2008). The forecasting of International Expo tourism using quantitative and qualitative techniques. *Tourism Management, 29*(6), 1084–1098. https://doi.org/10.1016/j.tourman.2008.02.007

Lei, F. (2024). Online recruitment for an online survey study: Our experience of dealing with fraudsters. *Applied Nursing Research, 80*, 151854. https://doi.org/10.1016/j.apnr.2024.151854

Liu, X., Cruz Rivera, S., Moher, D., Calvert, M. J., Denniston, A. K., Spirit, A. I., & Group, C.-A. W. (2020). Reporting guidelines for clinical trial reports for interventions involving artificial intelligence: The CONSORT-AI extension.

Lancet Digital Health, *2*(10), e537–e548. https://doi.org/10.1016/S2589-7500(20)30218-1

MacLennan, S., Kirkham, J., Lam, T. B. L., & Williamson, P. R. (2018). A randomized trial comparing three Delphi feedback strategies found no evidence of a difference in a setting with high initial agreement. *Journal of Clinical Epidemiology*, *93*, 1–8. https://doi.org/10.1016/j.jclinepi.2017.09.024

Manyara, A. M., Purvis, A., Ciani, O., Collins, G. S., & Taylor, R. S. (2024). Sample size in multistakeholder Delphi surveys: At what minimum sample size do replicability of results stabilize? *Journal of Clinical Epidemiology*, *174*, Article 111485. https://doi.org/10.1016/j.jclinepi.2024.111485

McAdam, M. K., Baldessarini, R. J., Murphy, A. L., & Gardner, D. M. (2023). Second international consensus study of antipsychotic dosing (ICSAD-2). *Journal of Psychopharmacology*, *37*(10), 982–991. https://doi.org/10.1177/02698811231205688

Meijering, J. V., & Tobi, H. (2016). The effect of controlled opinion feedback on Delphi features: Mixed messages from a real-world Delphi experiment. *Technological Forecasting and Social Change*, *103*, 166–173. https://doi.org/10.1016/j.techfore.2015.11.008

Meijering, J. V., & Tobi, H. (2018). The effects of feeding back experts' own initial ratings in Delphi studies: A randomized trial. *International Journal of Forecasting*, *34*(2), 216–224. https://doi.org/10.1016/j.ijforecast.2017.11.010

Minas, H., & Jorm, A. F. (2010). Where there is no evidence: Use of expert consensus methods to fill the evidence gap in low-income countries and cultural minorities. *International Journal of Mental Health Systems*, *4*, 33. https://doi.org/10.1186/1752-4458-4-33

Needham, D. M., Sepulveda, K. A., Dinglas, V. D., Chessare, C. M., Friedman, L. A., Bingham, C. O., III, & Turnbull, A. E. (2017). Core outcome measures for clinical research in acute respiratory failure survivors. An international modified Delphi consensus study. *American Journal of Respiratory and Critical Care Medicine*, *196*(9), 1122–1130. https://doi.org/10.1164/rccm.201702-0372OC

Niederberger, M., & Homberg, A. (2023). Argument-based QUalitative Analysis strategy (AQUA) for analyzing free-text responses in health sciences Delphi studies. *MethodsX*, *10*, 102156. https://doi.org/10.1016/j.mex.2023.102156

Niederberger, M., Schifano, J., Deckert, S., Hirt, J., Homberg, A., Köberich, S., Kuhn, R., Rommel, A., Sonnberger, M., Alam, M., Backman, C., Banno, M., Bartoszko, J., Bloomfield, F., Bober, M. B., Chalkoo, M., Chan, T. M., Chen, Y., Coscia, C., et al. (2024). Delphi studies in social and health sciences—Recommendations for an interdisciplinary standardized reporting (DELPHISTAR). Results of a Delphi study. *PLoS One*, *19*(8), Article e0304651. https://doi.org/10.1371/journal.pone.0304651

Niederberger, M., & Spranger, J. (2020). Delphi technique in health sciences: A map. *Frontiers in Public Health*, *8*, 457. https://doi.org/10.3389/fpubh.2020.00457

Nilsson, K., Andersson, G., Johansson, P., & Lundgren, J. (2023). Developing and designing an internet-based support and education program for patients awaiting kidney transplantation with deceased donors through: A Delphi study. *BMC Nephrology*, *24*(1), 311. https://doi.org/10.1186/s12882-023-03364-2

Nosek, B. A., Ebersole, C. R., DeHaven, A. C., & Mellor, D. T. (2018). The pre-registration revolution. *Proceedings of the National Academy of Sciences USA*, *115*(11), 2600–2606. https://doi.org/10.1073/pnas.1708274114

Parks, S., d'Angelo, C., & Gunashekar, S. (2018). *Citizen science: Generating ideas and exploring consensus*. The Healthcare Improvement Studies Institute.

Paul, C. L., Sanson-Fisher, R., Douglas, H. E., Clinton-McHarg, T., Williamson, A., & Barker, D. (2011). Cutting the research pie: A value-weighting approach to explore perceptions about psychosocial research priorities for adults with haematological cancers. *European Journal of Cancer Care*, *20*(3), 345–353. https://doi.org/10.1111/j.1365-2354.2010.01188.x

Quirke, F. A., Battin, M. R., Bernard, C., Biesty, L., Bloomfield, F. H., Daly, M., Finucane, E., Haas, D. M., Healy, P., Hurley, T., Koskei, S., Meher, S., Molloy, E. J., Niaz, M., Bhraonáin, E. N., Okaronon, C. O., Tabassum, F., Walker, K., Webbe, J. R. H., et al. (2023). Multi-Round versus Real-Time Delphi survey approach for achieving consensus in the COHESION core outcome set: A randomised trial. *Trials*, *24*(1), 461. https://doi.org/10.1186/s13063-023-07388-9

Radomski, T. R., Decker, A., Khodyakov, D., Thorpe, C. T., Hanlon, J. T., Roberts, M. S., Fine, M. J., & Gellad, W. F. (2022). Development of a metric to detect and decrease low-value prescribing in older adults. *JAMA Network Open*, *5*(2), e2148599. https://doi.org/10.1001/jamanetworkopen.2021.48599

Ramke, J., Evans, J. R., Habtamu, E., Mwangi, N., Silva, J. C., Swenor, B. K., Congdon, N., Faal, H. B., Foster, A., Friedman, D. S., Gichuhi, S., Jonas, J. B., Khaw, P. T., Kyari, F., Murthy, G. V. S., Wang, N., Wong, T. Y., Wormald, R., Yusufu, M., et al. (2022). Grand challenges in global eye health: A global prioritisation process using Delphi method. *Lancet Healthy Longevity*, *3*(1), e31–e41. https://doi.org/10.1016/s2666-7568(21)00302-0

Robinson, J., Thorn, P., McKay, S., Hemming, L., Battersby-Coulter, R., Cooper, C., Veresova, M., Li, A., Reavley, N., Rice, S., Lamblin, M., Pirkis, J., Reidenberg, D., Harrison, V., Skehan, J., & La Sala, L. (2023). #chatsafe 2.0. Updated guidelines to support young people to communicate safely online about self-harm and suicide: A Delphi expert consensus study. *PLoS One*, *18*(8), e0289494. https://doi.org/10.1371/journal.pone.0289494

Ross, A. M., Kelly, C. M., & Jorm, A. F. (2014). Re-development of mental health first aid guidelines for suicidal ideation and behaviour: A Delphi study. *BMC Psychiatry, 14*, 241. https://doi.org/10.1186/s12888-014-0241-8

Rowe, G., & Wright, G. (1996). The impact of task characteristics on the performance of structured group forecasting techniques. *International Journal of Forecasting, 12*(1), 73–89. https://doi.org/10.1016/0169-2070(95)00658-3

Rowe, G., Wright, G., & McColl, A. (2005). Judgment change during Delphi-like procedures: The role of majority influence, expertise, and confidence. *Technological Forecasting and Social Change, 72*(4), 377–399. https://doi.org/10.1016/j.techfore.2004.03.004

Sahle, B. W., Reavley, N. J., Morgan, A. J., Yap, M. B. H., Reupert, A., & Jorm, A. F. (2022). A Delphi study to identify intervention priorities to prevent the occurrence and reduce the impact of adverse childhood experiences. *The Australian and New Zealand Journal of Psychiatry, 56*(6), 686–694. https://doi.org/10.1177/00048674211025717

Saunders, H., Anderson, C., Feldman, F., Holroyd-Leduc, J., Jain, R., Liu, B., Macaulay, S., Marr, S., Silvius, J., Weldon, J., Bayoumi, A. M., Straus, S. E., Tricco, A. C., & Isaranuwatchai, W. (2023). Developing a fall prevention intervention economic model. *PLoS One, 18*(1 January), Article e0280572. https://doi.org/10.1371/journal.pone.0280572

Schifano, J., & Niederberger, M. (2025). How Delphi studies in the health sciences find consensus: A scoping review. *Systematic Reviews, 14*(1), 14. https://doi.org/10.1186/s13643-024-02738-3

Sii, S., Barton, K., Pasquale, L. R., Yamamoto, T., King, A. J., & Azuara-Blanco, A. (2018). Reporting harm in glaucoma surgical trials: Systematic review and a consensus-derived new classification system. *American Journal of Ophthalmology, 194*, 153–162. https://doi.org/10.1016/j.ajo.2018.07.014

Sinha, I. P., Smyth, R. L., & Williamson, P. R. (2011). Using the Delphi technique to determine which outcomes to measure in clinical trials: Recommendations for the future based on a systematic review of existing studies. *PLoS Medicine, 8*(1), e1000393. https://doi.org/10.1371/journal.pmed.1000393

Smelt, A. F., Louter, M. A., Kies, D. A., Blom, J. W., Terwindt, G. M., van der Heijden, G. J., De Gucht, V., Ferrari, M. D., & Assendelft, W. J. (2014). What do patients consider to be the most important outcomes for effectiveness studies on migraine treatment? Results of a Delphi study. *PLoS One, 9*(6), e98933. https://doi.org/10.1371/journal.pone.0098933

Spranger, J., Homberg, A., Sonnberger, M., & Niederberger, M. (2022). Reporting guidelines for Delphi techniques in health sciences: A methodological review. *Zeitschrift für Evidenz, Fortbildung und Qualität im Gesundheitswesen, 172*, 1–11. https://doi.org/10.1016/j.zefq.2022.04.025

Swedo, S. E., Baguley, D. M., Denys, D., Dixon, L. J., Erfanian, M., Fioretti, A., Jastreboff, P. J., Kumar, S., Rosenthal, M. Z., Rouw, R., Schiller, D., Simner,

J., Storch, E. A., Taylor, S., Werff, K. R. V., Altimus, C. M., & Raver, S. M. (2022). Consensus definition of misophonia: A Delphi study. *Frontiers in Neuroscience, 16*, 841816. https://doi.org/10.3389/fnins.2022.841816

Ter Veer, E., van Rijssen, L. B., Besselink, M. G., Mali, R. M. A., Berlin, J. D., Boeck, S., Bonnetain, F., Chau, I., Conroy, T., Van Cutsem, E., Deplanque, G., Friess, H., Glimelius, B., Goldstein, D., Herrmann, R., Labianca, R., Van Laethem, J. L., Macarulla, T., van der Meer, J. H. M., et al. (2018). Consensus statement on mandatory measurements in pancreatic cancer trials (COMM-PACT) for systemic treatment of unresectable disease. *Lancet Oncology, 19*(3), e151–e160. https://doi.org/10.1016/s1470-2045(18)30098-6

Thoomes, E., Falla, D., Cleland, J. A., Fernández-de-Las-Peñas, C., Gallina, A., & de Graaf, M. (2023). Conservative management for lumbar radiculopathy based on the stage of the disorder: A Delphi study. *Disability and Rehabilitation, 45*(21), 3539–3548. https://doi.org/10.1080/09638288.2022.2130448

Toulany, A., Khodyakov, D., Mooney, S., Stromquist, L., Bailey, K., Barber, C. E., Batthish, M., Cleverley, K., Dimitropoulos, G., Gorter, J. W., Grahovac, D., Grimes, R., Guttman, B., Hébert, M. L., John, T., Lo, L., Luong, D., MacGregor, L., Mukerji, G., et al. (2024). Quality indicators for transition from pediatric to adult care for youth with chronic conditions: Proposal for an online modified Delphi study. *JMIR Research Protocols, 13*, e60860. https://doi.org/10.2196/60860

Turnbull, A. E., Dinglas, V. D., Friedman, L. A., Chessare, C. M., Sepúlveda, K. A., Bingham, C. O., III, & Needham, D. M. (2018). A survey of Delphi panelists after core outcome set development revealed positive feedback and methods to facilitate panel member participation. *Journal of Clinical Epidemiology, 102*, 99–106. https://doi.org/10.1016/j.jclinepi.2018.06.007

Waggoner, J., Carline, J. D., & Durning, S. J. (2016). Is there a consensus on consensus methodology? Descriptions and recommendations for future consensus research. *Academic Medicine, 91*(5), 663–668. https://doi.org/10.1097/acm.0000000000001092

Yap, M. B. H., Fowler, M., Reavley, N., & Jorm, A. F. (2015). Parenting strategies for reducing the risk of childhood depression and anxiety disorders: A Delphi consensus study. *Journal of Affective Disorders, 183*, 330–338. https://doi.org/10.1016/j.jad.2015.05.031

Yap, M. B. H., & Jorm, A. F. (2015). Parental factors associated with childhood anxiety, depression, and internalizing problems: A systematic review and meta-analysis. *Journal of Affective Disorders, 175*, 424–440. https://doi.org/10.1016/j.jad.2015.01.050

Yap, M. B., Pilkington, P. D., Ryan, S. M., Kelly, C. M., & Jorm, A. F. (2014). Parenting strategies for reducing the risk of adolescent depression and anxiety disorders: A Delphi consensus study. *Journal of Affective Disorders, 156*, 67–75. https://doi.org/10.1016/j.jad.2013.11.017

Yu, M., Keel, S., Mariotti, S., Mills, J., & Müller, A. (2023). Development of the WHO eye care competency framework. *Human Resources for Health, 21*(1), 46. https://doi.org/10.1186/s12960-023-00834-4

Zickafoose, A., Lu, P., & Baker, M. (2022). Forecasting food innovations with a Delphi study. *Food, 11*(22). https://doi.org/10.3390/foods11223723

CHAPTER 5

Implementing the Findings of Delphi Studies in Practice

Abstract This chapter argues that the publication of a Delphi study is unlikely in itself to lead to any practical change and that implementation needs to be considered when planning a study. Useful ways to facilitate implementation include producing a program logic that makes clear the link between the Delphi study and desired outcomes, using the Delphi method to establish consensus on implementation strategies, including knowledge users as experts and collaborating with an implementing organization.

Keywords Delphi method • Implementation • Experts • Knowledge user • Collaboration

The message of this chapter is that publishing a Delphi study will not in itself change anything. If a Delphi study is to produce change, there needs to be a plan from the start about how this will be achieved.

Chapter 1 noted that many uses of the Delphi method fall in one of the following general categories:

• Making judgements on facts where the evidence is imperfect
• Setting methodological standards
• Making predictions

© The Author(s) 2025
A. Jorm, *Using the Delphi Method to Establish Expert Consensus*,
Advancing Methods for Interdisciplinarity in the Social Sciences,
https://doi.org/10.1007/978-981-96-8357-4_5

95

- Defining foundational concepts
- Determining collective values
- Improving professional practice
- Improving policy

Two of these uses ("Improving professional practice" and "Improving policy") directly state a broader practical aim. However, many of the other uses also imply some change to practice. For example, researchers who carry out a Delphi study for "Setting methodological standards" or "Defining foundational concepts" will want to see these standards or concepts adopted by others. Similarly, "Determining collective values" will generally entail being guided by these values for some practical purpose, such as allocation of resources. In such cases, the aim of a Delphi study can be seen as a stepping stone to some broader practical aim of producing change in how things are done.

Although a change in practice might be an implicit longer-term aim of many Delphi studies, publications reporting these studies rarely discuss how this might be achieved. Indeed, neither of the two major standards for reporting Delphi studies, ACCORD and DELPHISTAR (see Chap. 4), contain any items concerned with discussing the implementation of Delphi findings into practice.

This gap between the immediate aim of the research and the longer-term aim of changing practice is not confined to Delphi studies. In the area of health sciences, it has often been noted that findings on interventions from randomized controlled trials are not routinely implemented in practice. Implementation of intervention findings may eventually occur through spontaneous diffusion, but there can be a lag of decades and the implementation may be incomplete or patchy. Concern about this gap has given rise to the discipline of "implementation science", which has been defined as "the scientific study of methods to promote the systematic uptake of research findings and other evidence-based practices into routine practice, and, hence, to improve the quality and effectiveness of health services and care" (Eccles & Mittman, 2006, p. 1). While the field of implementation science has been concerned with improving healthcare, a similar research-to-practice gap is common in other areas of research.

Although the research-to-practice gap occurs with research methodologies other than randomized controlled trials, it has received very little attention in relation to consensus methodologies such as Delphi. My own reading of the implementation science literature reveals very little that might be useful guidance for Delphi researchers. In the absence of an existing implementation science for consensus studies, I offer some general principles based on the scant literature and some case examples of successful implementation. Such principles should arguably also be based on expert consensus, but in the absence of this I offer the suggestions below as a starting point for filling the gap.

1   IMPLEMENTATION NEEDS TO BE CONSIDERED WHEN PLANNING A DELPHI STUDY

The most important recommendation is to include implementation when planning a Delphi study. It is rare to find a report on a Delphi study that has any mention of implementation beyond a general conclusion that someone should act on the findings. Below are some typical examples.

> Future research should focus on pragmatic application of these recommendations to ensure growing use of PROs [patient-reported outcomes] into clinical practice. (Mazariego et al., 2022, p. 41)
>
> Our findings should be used in the planning and funding of PPI [patient and public involvement] in clinical trials to help focus research efforts and minimize waste. (Kearney et al., 2017, p. 1401)
>
> We strongly urge researchers, public servants and non-government organisations to use this definition to inform advocacy, research, strategy development and evaluation frameworks. (Gallegos et al., 2023, p. 1995).

However, there are a small minority of Delphi study publications that report that the authors have planned for the implementation of their findings. Example 5.1 illustrates this with a Delphi study to develop guidelines on how adolescents should best support their peers with mental health problems. In this case, the findings were used to develop the curriculum content of a school-based training program that was subsequently evaluated in trials.

Example 5.1 Implementation Plan for a Delphi Study on How Adolescents Should Provide Mental Health First Aid to Peers

Ross et al. (2012) carried out a Delphi study to establish consensus on key messages for adolescents on providing basic mental health first aid to peers. The experts were youth mental health consumer advocates or Mental Health First Aid (MHFA) instructors delivering courses for adults to assist the youth. There was a consensus on 78 key messages for junior adolescents and 81 for senior adolescents. In this study, the authors outlined the following implementation plan.

> The statements that were endorsed in this study will be used to guide the development of the curriculum for a basic MHFA course for adolescents, tailored to their developmental levels. This course will complement the existing Youth MHFA course, which teaches adult members of the public to assist adolescents in getting help. Although the current research does not provide direct evidence that adolescents employing the endorsed messages will improve the outcome for peers with a mental health problem, once implemented the course will be evaluated to determine the impact on first aid behaviours and mental health outcomes. Such an evaluation will investigate whether the basic MHFA course is effective in changing the knowledge, attitudes and behaviours of adolescents towards their peers with mental health problems. (p. 237)

Subsequently, the Delphi findings were used to develop a Teen Mental Health First Aid Action Plan and a course to teach it. After development, the course was evaluated in an uncontrolled trial and a randomized controlled trial, showing the hypothesized changes (Hart et al., 2016, 2022) and it has been disseminated in a number of countries (https://mspgh.unimelb.edu.au/research-groups/centre-for-health-equity/equity-and-mental-health/teen-mental-health-first-aid).

Example 5.2 describes a different type of implementation plan, involving a Delphi study to update the guidelines for reporting systematic reviews.

Example 5.2 Implementation Plan for a Delphi Study to Update Guidelines for Reporting Systematic Reviews
The Preferred Reporting Items for Systematic Reviews and Meta-Analyses (PRISMA) statement was first published in 2009 to help authors of systematic reviews to improve the quality of reporting in their publications. To update the statement, Page et al. (2018) developed a study protocol which involved a new review of the literature and a Delphi study. The Delphi study involved asking a diverse panel of experts to assess the appropriateness of keeping existing PRISMA items and adding potential new ones. The study protocol included the following implementation strategies.

We plan to submit a paper presenting the updated PRISMA checklist, and a revised explanation and elaboration document, to an open-access journal. The explanation and elaboration document will provide detailed guidance on each item along with exemplars of item reporting from published SRs [systematic reviews]. We will revise the existing PRISMA statement website (http://www.prisma-statement. org/), so that it includes the latest versions of the checklist and explanation and elaboration papers. We will raise awareness of the updated guideline via presentations and workshops at relevant conferences focused on SR methodology, health technology assessment and evidence-based medicine … twitter accounts, and in a series of international webinars. … We plan to draft standardized text to include in instructions for authors and peer reviewers, create an editable checklist for authors to submit to journals, and a template PRISMA flow diagram. We will also consider developing an online writing tool based on the updated PRISMA checklist. … We will also liaise with the developers of StatReviewer software, which performs an automated audit of the statistical and reporting integrity of scientific manuscripts. (p. 12)

The Delphi study and an associated in-person meeting resulted in substantial changes to the 2009 statement, which was published as the updated PRISMA 2020 statement (Page et al., 2021) and is available at the PRISMA (www.prisma-statement.org/prisma-2020) and Equator Network (www.equator-network.org/reporting-guidelines/prisma) websites.

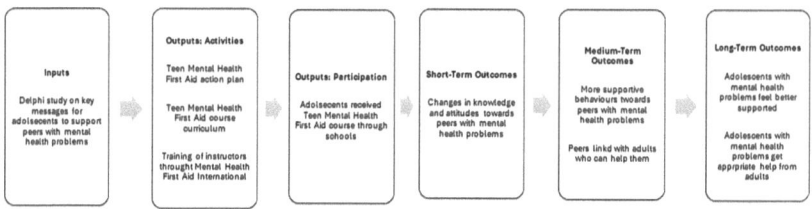

Fig. 5.1 A program logic to improve peer social support and appropriate help from adults for adolescents with mental health problems

A useful aid for thinking about implementation is a program logic model. This has been defined as "a schematic representation that describes how a program is intended to work by linking activities with outputs, intermediate impacts and longer term outcomes" (Centre for Epidemiology and Evidence, 2017, p. 4). The program logic is a causal model showing how various inputs (such as a Delphi study) can be linked through a series of steps to a longer-term practical goal. There are various ways of representing a program logic. The simplest approach is a pipeline program logic model like that shown in Fig. 5.1, which presents implementation as a series of linear steps. This program logic model is based on the Delphi study described in Example 5.1.

More complex program logic models are also possible. A publication by the Centre for Epidemiology and Evidence (Centre for Epidemiology and Evidence, 2017) provides a clear explanation of how to develop a program logic model and the various options available for representing them.

2   USING DELPHI TO INFORM
IMPLEMENTATION STRATEGIES

Examples 5.1 and 5.2 above show how Delphi researchers can come up with an implementation plan for their findings. However, it is also possible to use the Delphi method as a way of getting a broader consensus on the best approach to implementation. Such a consensus could cover issues like barriers and facilitators to implementation, implementation strategies, organizational support needed, and resources required.

Example 5.3 illustrates how a Delphi study was used to determine the important factors that need to be addressed for the implementation of youth care guidelines, while Example 5.4 shows how Delphi was used to come up with a strategy to overcome an important barrier to the implementation of a health intervention, namely a lack of health insurance coverage.

Example 5.3 Use of Delphi to Develop Consensus on Important Determinants for the Implementation of Youth Care Guidelines
The Netherlands has several practice guidelines related to the care of youth, covering issues such as youth psychosocial problems, child abuse, domestic violence, and parenting problems. However, the availability of guidelines does not mean that they will be implemented by practitioners. Even when professionals are aware of the guidelines, many do not implement them in practice. To inform more effective implementation, Dubbeldeman et al. (2024) carried out a Delphi study to get a consensus of experts on factors that are relevant to the implementation of guidelines and that are changeable in practice. They recruited 14 Dutch panellists with expertise in guideline implementation who rated possible determinants identified from a literature review and some available data. The experts rated the most important determinants of implementation to be guideline promotion, mandatory education, knowledge regarding guideline use, presence of an implementation leader, poor management support, and lack of communication skills. For each of these determinants, the panellists developed hypotheses for how the determinant could be addressed. For example, for the determinant of "communication skills", the implementation hypothesis was: "Practice communication skills during, for example, role-play to build habits and skills. Skills training should be provided in an ongoing way so that professionals can repeatedly practice their communication skills and new employees can join training sessions" (p. 14). It is ironic, however, that the authors of this study did not present a plan for implementing these implementation hypotheses to improve the use of youth care guidelines. They simply concluded that: "This study offers a set of hypotheses that could help organisations, policymakers, and professionals guide the implementation process of youth care guidelines to ultimately improve implementation outcomes" (p. 1). No further action has been reported.

Example 5.4 Use of Delphi to Develop Consensus on How to Get Mindfulness-Based Stress Reduction Covered by Health Insurance

Mindfulness-Based Stress Reduction (MBSR) is a group treatment program based on mindfulness meditation which is used for stress management and to treat a variety of mental and physical conditions. Although there is evidence for its effectiveness from randomized controlled trials, MBSR is not generally covered by health insurance in the United States, limiting its uptake. To deal with this gap, Frank et al. (2024) carried out a Delphi study to get consensus on if and how MBSR should be covered by health insurance, the barriers and facilitators involved, and what is the highest priority evidence needed to inform health insurance coverage decisions. There were 26 panellists, including health insurers, healthcare administrators, policymakers, clinicians, MBSR instructors and MBSR students. Items to be rated were based on initial qualitative interviews. The experts endorsed the importance of the evidence that MBSR is effective and not harmful. Facilitators included the potential of the treatment to be used for common mental health and psychosomatic problems. Barriers included that it is not a medical treatment and patient barriers to attendance. The highest priority research was on what conditions the treatment is effective for and its impact on stress. The Delphi findings were discussed by an advisory board that selected five priority areas for action in advancing MBSR insurance coverage.

3   Including Knowledge Users as Expert Panellists

Implementation may be more likely if the end users of research are actively involved in the research process. This is easier for Delphi studies than for other methodologies, as these studies can include potential knowledge users as panellists. Such knowledge users could be either implementors of the knowledge such as practitioners and administrators, or recipients of implementation such as patients, caregivers, customers and consumers.

Example 5.5 illustrates the involvement of knowledge users in a study to promote the implementation of guidelines on optimal amounts of movement for good health. In this case the knowledge users were primary health providers, and they were involved as Delphi panellists to develop a toolkit that providers could use with their patients. The panellists were also involved in the planning of an implementation strategy in collaboration with the researchers.

Example 5.5 Use of the Delphi Method to Promote the Use of Canadian Movement Guidelines by Primary Care Providers
Canada has national 24-hour movement guidelines which recommend optimal amounts of physical activity, sedentary behaviour, and sleep for good health. Primary care providers have a potentially important role in promoting the guideline recommendations to their patients but may lack the time or be uncertain about how to do this. To help primary care providers in this role, Morgan et al. (2025) used a modified Delphi method to develop a toolkit for how to initiate discussions with patients about their activity levels and promote behaviour change. The authors produced a draft version of "The Whole Day Matters Tool" based on a literature review, usability study, and a counselling framework. They then carried out the Delphi study with a diverse panel of 20 primary care providers, including physicians, nurses, dietitians, psychologists, and pharmacists. The panellists were asked to rate the utility, acceptability, and understandability of each of the toolkit components and provide comments. These ratings and comments were used to revise the toolkit to its final form. The researchers also worked with the panellists on a dissemination strategy, which involved producing a short video that was posted on professional association websites, mass emails to user groups, and posting messages on a range of social media. To evaluate uptake, they recorded the number of downloads of the toolkit, finding that it had been downloaded over 1000 times in the 4 months following release.

---

4 Collaborating with an Implementing Organization

Another implementation strategy is for Delphi researchers to form a collaboration with an implementing organization. The ideal situation might be where the implementing organization commissions the researchers to carry out the Delphi research project, so that the longer-term implementation goal is at the forefront. However, it is also possible for researchers to take the initiative in partnering with the implementing organization or to form such an organization themselves.

Two examples are given below. Example 5.6 shows how two research networks partnered with each other and with an advocacy organization (the Alzheimer's Association) to implement the findings from a Delphi study on the analysis of MRI data that is important to the diagnosis of Alzheimer's disease. The implementation activities included creating a website promoting the method developed and producing a training program in its use.

Example 5.6 Delphi Study Carried out by Implementing Organizations to Develop a Harmonized Protocol for Segmentation of the Hippocampus on Magnetic Resonance Imaging

Atrophy of the hippocampus is used as a feature to support the clinical diagnosis of Alzheimer's disease. This atrophy can be assessed using magnetic resonance imaging but requires that an expert manually trace the boundaries of the hippocampus. Various protocols have been developed for tracing these boundaries, but these do not always agree in the details, making it difficult to compare results from different laboratories. To produce a harmonized protocol that could be used across laboratories, Boccardi et al. (2015) carried out a Delphi study with experts internationally. To coordinate this work, the project was carried out as a collaboration between two large research networks, the European Alzheimer's Disease Consortium (EADC) and the Alzheimer's Disease Neuroimaging Initiative (ADNI), supported by the Alzheimer's Association. This work resulted in the EADC-ADNI Harmonized Protocol for segmentation of the hippocampus. A website was created for the dissemination of the Harmonized Protocol (www.hippocampal-protocol.net/SOPs/index.php), which included a web training platform allowing tracers to learn and qualify for hippocampal segmentation based on the Protocol.

The second example involves the use of Delphi studies to produce mental health first aid guidelines which were then used by the implementation partner Mental Health First Aid International to inform the curriculum content of training courses that have had global dissemination (Example 5.7).

Example 5.7 Implementation of Delphi-Based Mental Health First Aid Guidelines by Mental Health First Aid International
Mental Health First Aid training courses teach members of the public how to support a person developing a mental health problem, experiencing a worsening of an existing mental health problem or in a mental health crisis (e.g. they are suicidal or self-harming). To inform the content of Mental Health First Aid training, a series of Delphi studies have been carried out with mental health professionals and people with lived experience as experts to develop mental health first aid guidelines covering a range of mental health problems and crises (Jorm & Ross, 2018). Guidelines have also been developed using cultural expertise on how to assist people from special groups (e.g. Indigenous Australians; refugees and immigrants; lesbian, gay, bisexual and transgender people). These studies have been carried out as a partnership between academic researchers and staff of Mental Health First Aid International, an Australian-based not-for-profit organization which develops and disseminates Mental Health First Aid courses (Jorm et al., 2019). This organization has managed to spread the training to 29 countries, with more than 8 million people trained worldwide as Mental Health First Aiders (https://mhfainternational.org/international-mental-health-first-aid-programs/).

Successful implementation is more likely when an implementing organization is the initiator of Delphi research, as it will be directly informing their needs. Example 5.8 illustrates how a government-funded partnership between two hospital services used a Delphi consensus study to develop practical tools for infection control.

Example 5.8 Development of Practical Tools to Enhance Infection Prevention and Control in Healthcare Facilities

The US Centers for Disease Control and Prevention (CDC) funded a partnership between the New York City Health & Hospitals/ Bellevue and the Massachusetts General Hospital to develop practical tools for infection control in healthcare facilities. As part of this work, a Delphi study was carried out on how to prepare the infrastructure of acute care facilities for high-consequence infectious diseases, such as viral haemorrhagic fevers caused by Ebola, Lassa and Marburg viruses (Chan et al., 2024). A group of 29 professionals with expertise in infection control endorsed 36 statements covering patient physical space, doors and windows, air handling, electrical and plumbing, and soiled utility rooms and waste management. To disseminate this consensus, an interactive online tool was created and hosted on the project website. This tool shows images of a mocked-up facility incorporating the consensus features. Users can click on any infrastructure feature to see the consensus statement that it represents (https://mypact.us/resources/high-level-isolation-room-interactive-tool-patient-room-overview/).

5   Conclusion

The basic message of this chapter is that implementation of Delphi findings is unlikely unless there is a plan for it from the beginning. In the absence of an implementation science for consensus studies, the chapter offers a range of examples of successful implementation which the reader can use as a source of ideas for their own implementation goals.

References

Boccardi, M., Bocchetta, M., Apostolova, L. G., Barnes, J., Bartzokis, G., Corbetta, G., DeCarli, C., deToledo-Morrell, L., Firbank, M., Ganzola, R., Gerritsen, L., Henneman, W., Killiany, R. J., Malykhin, N., Pasqualetti, P., Pruessner, J. C., Redolfi, A., Robitaille, N., Soininen, H., et al. (2015). Delphi definition of the EADC-ADNI harmonized protocol for hippocampal segmentation on magnetic resonance. *Alzheimer's & Dementia, 11*(2), 126–138. https://doi.org/10.1016/j.jalz.2014.02.009

Centre for Epidemiology and Evidence. (2017). *Developing and using program logic: A guide.* Ministry of Health.

Chan, J., Searle, E. F., Khodyakov, D., Denson, L., Echeverri, A., Browne, E. M., Chiarelli, Y., Dickey, L. L., Erickson, D. S., Flannery, J., Kaplan, L. J., Markovitz, S., Popescu, S. V., & Shenoy, E. S. (2024). They will come, you must build it: A modified Delphi process applied to preparing acute care facilities infrastructure for high-consequence infectious diseases. *Health Security, 22*(5), 384–393. https://doi.org/10.1089/hs.2024.0013

Dubbeldeman, E. M., van der Kleij, R. M. J. J., Brakema, E. A., & Crone, M. R. (2024). Expert consensus on multilevel implementation hypotheses to promote the uptake of youth care guidelines: A Delphi study. *Health Research Policy and Systems, 22*(1), 89. https://doi.org/10.1186/s12961-024-01167-x

Eccles, M. P., & Mittman, B. S. (2006). Welcome to Implementation Science. *Implementation Science, 1*(1), 1. https://doi.org/10.1186/1748-5908-1-1

Frank, H. E., Albanese, A., Sun, S., Saadeh, F., Johnson, B. T., Elwy, A. R., & Loucks, E. B. (2024). Mindfulness-based stress reduction health insurance coverage: If, how, and when? An integrated knowledge translation (iKT) Delphi key informant analysis. *Mindfulness, 15*(5), 1220–1233. https://doi.org/10.1007/s12671-024-02366-x

Gallegos, D., Booth, S., Pollard, C. M., Chilton, M., & Kleve, S. (2023). Food security definition, measures and advocacy priorities in high-income countries: A Delphi consensus study. *Public Health Nutrition, 26*(10), 1986–1996. https://doi.org/10.1017/s1368980023000915

Hart, L. M., Mason, R. J., Kelly, C. M., Cvetkovski, S., & Jorm, A. F. (2016). 'teen Mental Health First Aid': A description of the program and an initial evaluation. *International Journal of Mental Health Systems, 10*, 3. https://doi.org/10.1186/s13033-016-0034-1

Hart, L. M., Morgan, A. J., Rossetto, A., Kelly, C. M., Gregg, K., Gross, M., Johnson, C., & Jorm, A. F. (2022). teen Mental Health First Aid: 12-month outcomes from a cluster crossover randomized controlled trial evaluation of a universal program to help adolescents better support peers with a mental health problem. *BMC Public Health, 22*(1), 1159. https://doi.org/10.1186/s12889-022-13554-6

Jorm, A. F., Kitchener, B. A., & Reavley, N. J. (2019). Mental Health First Aid training: Lessons learned from the global spread of a community education program. *World Psychiatry, 18*(2), 142–143. https://doi.org/10.1002/wps.20621

Jorm, A. F., & Ross, A. M. (2018). Guidelines for the public on how to provide mental health first aid: Narrative review. *BJPsych Open, 4*(6), 427–440. https://doi.org/10.1192/bjo.2018.58

Kearney, A., Williamson, P., Young, B., Bagley, H., Gamble, C., Denegri, S., Muir, D., Simon, N. A., Thomas, S., Elliot, J. T., Bulbeck, H., Crocker, J. C., Planner,

C., Vale, C., Clarke, M., Sprosen, T., & Woolfall, K. (2017). Priorities for methodological research on patient and public involvement in clinical trials: A modified Delphi process. *Health Expectations, 20*(6), 1401–1410. https://doi. org/10.1111/hex.12583

Mazariego, C., Jefford, M., Chan, R. J., Roberts, N., Millar, L., Anazodo, A., Hayes, S., Brown, B., Saunders, C., Webber, K., Vardy, J., Girgis, A., & Koczwara, B. (2022). Priority recommendations for the implementation of patient-reported outcomes in clinical cancer care: A Delphi study. *Journal of Cancer Survivorship, 16*(1), 33–43. https://doi.org/10.1007/s11764-021-01135-2

Morgan, T. L., Fortier, M. S., Jain, R., Lane, K. N., Maclaren, K., McFadden, T., Prorok, J., Robison, J., Weston, Z. J., & Tomasone, J. R. (2025). Development of the Whole Day Matters Toolkit for Primary Care: A consensus-building study to mobilize national public health guidelines in practice. *Health Promotion and Chronic Disease Prevention in Canada, 45*(1), 1–19. https://doi. org/10.24095/hpcdp.45.1.01

Page, M. J., McKenzie, J. E., Bossuyt, P. M., Boutron, I., Hoffmann, T., Mulrow, C. D., Shamseer, L., & Moher, D. (2018). *Updating the PRISMA reporting guideline for systematic reviews and meta-analyses: Study protocol.* Monash University School of Public Health and Preventive Medicine.

Page, M. J., McKenzie, J. E., Bossuyt, P. M., Boutron, I., Hoffmann, T. C., Mulrow, C. D., Shamseer, L., Tetzlaff, J. M., Akl, E. A., Brennan, S. E., Chou, R., Glanville, J., Grimshaw, J. M., Hrobjartsson, A., Lalu, M. M., Li, T., Loder, E. W., Mayo-Wilson, E., McDonald, S., et al. (2021). The PRISMA 2020 statement: An updated guideline for reporting systematic reviews. *BMJ, 372*, n71. https://doi.org/10.1136/bmj.n71

Ross, A. M., Hart, L. M., Jorm, A. F., Kelly, C. M., & Kitchener, B. A. (2012). Development of key messages for adolescents on providing basic mental health first aid to peers: A Delphi consensus study. *Early Intervention in Psychiatry, 6*(3), 229–238. https://doi.org/10.1111/j.1751-7893.2011.00331.x

References

Aengenheyster, S., Cuhls, K., Gerhold, L., Heiskanen-Schüttler, M., Huck, J., & Muszynska, M. (2017). Real-time Delphi in practice—A comparative analysis of existing software-based tools. *Technological Forecasting and Social Change, 118*, 15–27. https://doi.org/10.1016/j.techfore.2017.01.023

Ahmed, I., & Metcalfe, A. (2024). Research priorities of members of the British Association for Surgery of the Knee. *Bone and Joint Journal, 106 B*(7), 662–668. https://doi.org/10.1302/0301-620X.106B7.BJJ-2023-0691.R1

Akins, R. B., Tolson, H., & Cole, B. R. (2005). Stability of response characteristics of a Delphi panel: Application of bootstrap data expansion. *BMC Medical Research Methodology, 5*, 37. https://doi.org/10.1186/1471-2288-5-37

Amaratunge, S., Harrison, M., Clifford, R., Seubert, L., Page, A., & Bond, C. (2021). Developing a checklist for reporting research using simulated patient methodology (CRiSP): A consensus study. *International Journal of Pharmacy Practice, 29*(3), 218–227. https://doi.org/10.1093/ijpp/riaa002

American Psychological Association. (1966). *Standards for educational and psychological tests and manuals*. APA.

Andrews, G., Bell, C., Boyce, P., Gale, C., Lampe, L., Marwat, O., Rapee, R., & Wilkins, G. (2018). Royal Australian and New Zealand College of Psychiatrists clinical practice guidelines for the treatment of panic disorder, social anxiety disorder and generalised anxiety disorder. *Australian and New Zealand Journal of Psychiatry, 52*, 1109–1172. https://doi.org/10.1177/0004867418799453

Arechar, A. A., Allen, J., Berinsky, A. J., Cole, R., Epstein, Z., Garimella, K., Gully, A., Lu, J. G., Ross, R. M., Stagnaro, M. N., Zhang, Y., Pennycook, G., &

© The Author(s) 2025
A. Jorm, *Using the Delphi Method to Establish Expert Consensus*,
Advancing Methods for Interdisciplinarity in the Social Sciences,
https://doi.org/10.1007/978-981-96-8357-4

Rand, D. G. (2023). Understanding and combatting misinformation across 16 countries on six continents. *Nature Human Behaviour, 7*(9), 1502–1513. https://doi.org/10.1038/s41562-023-01641-6

Baâdoudi, F., Picavet, S. H. S. J., Hildrink, H. B. M., Hendrikx, R., Rijken, M., & de Bruin, S. R. (2023). Are older people worse off in 2040 regarding health and resources to deal with it? – Future developments in complex health problems and in the availability of resources to manage health problems in the Netherlands [Article]. *Frontiers in Public Health, 11*, Article 942526. https://doi.org/10.3389/fpubh.2023.942526

Bailey, E., Bellairs-Walsh, I., Reavley, N., Gooding, P., Hetrick, S., Rice, S., Boland, A., & Robinson, J. (2024). Best practice for integrating digital interventions into clinical care for young people at risk of suicide: A Delphi study. *BMC Psychiatry, 24*(1), 71. https://doi.org/10.1186/s12888-023-05448-7

Baron, R. S. (2005). So right it's wrong: Groupthink and the ubiquitous nature of polarized group decision making. *Advances in Experimental Social Psychology, 37*, 219–253. https://doi.org/10.1016/S0065-2601(05)37004-3

Barrington, H., Young, B., & Williamson, P. R. (2021). Patient participation in Delphi surveys to develop core outcome sets: Systematic review. *BMJ Open, 11*(9), e051066. https://doi.org/10.1136/bmjopen-2021-051066

Barrios, M., Guilera, G., Nuño, L., & Gómez-Benito, J. (2021). Consensus in the delphi method: What makes a decision change? *Technological Forecasting and Social Change, 163*, 120484. https://doi.org/10.1016/j.techfore.2020.120484

Becker, J., Brackbill, D., & Centola, D. (2017). Network dynamics of social influence in the wisdom of crowds. *Proceedings of the National Academy of Sciences USA, 114*(26), E5070–E5076. https://doi.org/10.1073/pnas.1615978114

Belton, I., Wright, G., Sissons, A., Bolger, F., Crawford, M. M., Hamlin, I., Taylor Browne Lūka, C., & Vasilichi, A. (2021). Delphi with feedback of rationales: How large can a Delphi group be such that participants are not overloaded, de-motivated, or disengaged? *Technological Forecasting and Social Change, 170*, 120897. https://doi.org/10.1016/j.techfore.2021.120897

Berk, L., Jorm, A. F., Kelly, C. M., Dodd, S., & Berk, M. (2011). Development of guidelines for caregivers of people with bipolar disorder: A Delphi expert consensus study. *Bipolar Disorders, 13*(5–6), 556–570. https://doi.org/10.1111/j.1399-5618.2011.00942.x

Biddle, L., Rifkin-Zybutz, R., Derges, J., Turner, N., Bould, H., Sedgewick, F., Gooberman-Hill, R., Moran, P., & Linton, M. J. (2022). Developing good practice indicators to assist mental health practitioners to converse with young people about their online activities and impact on mental health: A two-panel mixed-methods Delphi study. *BMC Psychiatry, 22*(1), 485. https://doi.org/10.1186/s12888-022-04093-w

Boccardi, M., Bocchetta, M., Apostolova, L. G., Barnes, J., Bartzokis, G., Corbetta, G., DeCarli, C., deToledo-Morrell, L., Firbank, M., Ganzola, R., Gerritsen, L., Henneman, W., Killiany, R. J., Malykhin, N., Pasqualetti, P., Pruessner, J. C., Redolfi, A., Robitaille, N., Soininen, H., et al. (2015). Delphi definition of the EADC-ADNI harmonized protocol for hippocampal segmentation on magnetic resonance. *Alzheimer's & Dementia, 11*(2), 126–138. https://doi.org/10.1016/j.jalz.2014.02.009

Boel, A., Navarro-Compán, V., Landewé, R., & van der Heijde, D. (2021). Two different invitation approaches for consecutive rounds of a Delphi survey led to comparable final outcome. *Journal of Clinical Epidemiology, 129*, 31–39. https://doi.org/10.1016/j.jclinepi.2020.09.034

Bolger, F., Stranieri, A., Wright, G., & Yearwood, J. (2011). Does the Delphi process lead to increased accuracy in group-based judgmental forecasts or does it simply induce consensus amongst judgmental forecasters? *Technological Forecasting and Social Change, 78*(9), 1671–1680. https://doi.org/10.1016/j.techfore.2011.06.002

Boulkedid, R., Abdoul, H., Loustau, M., Sibony, O., & Alberti, C. (2011). Using and reporting the Delphi method for selecting healthcare quality indicators: A systematic review. *PLoS One, 6*(6), e20476. https://doi.org/10.1371/journal.pone.0020476

Brookes, S. T., Macefield, R. C., Williamson, P. R., McNair, A. G., Potter, S., Blencowe, N. S., Strong, S., & Blazeby, J. M. (2016). Three nested randomized controlled trials of peer-only or multiple stakeholder group feedback within Delphi surveys during core outcome and information set development. *Trials, 17*(1), 409. https://doi.org/10.1186/s13063-016-1479-x

Brown, B., Cochran, S., & Dalkey, N. (1969). *The DELPHI method, II: Structure of experitments*. RAND.

Burgman, M. A. (2016). *Trusting judgements: How to get the best out of experts*. Cambridge University Press.

Burns, P. B., Rohrich, R. J., & Chung, K. C. (2011). The levels of evidence and their role in evidence-based medicine. *Plastic and Reconstructive Surgery, 128*(1), 305–310. https://doi.org/10.1097/PRS.0b013e318219c171

Camerer, C. F., Dreber, A., Forsell, E., Ho, T. H., Huber, J., Johannesson, M., Kirchler, M., Almenberg, J., Altmejd, A., Chan, T., Heikensten, E., Holzmeister, F., Imai, T., Isaksson, S., Nave, G., Pfeiffer, T., Razen, M., & Wu, H. (2016). Evaluating replicability of laboratory experiments in economics. *Science, 351*, 1433–1436. https://doi.org/10.1126/science.aaf0

Campbell, S. M., Hann, M., Roland, M. O., Quayle, J. A., & Shekelle, P. G. (1999). The effect of panel membership and feedback on ratings in a two-round Delphi survey: Results of a randomized controlled trial. *Medical Care, 37*(9), 964–968. https://doi.org/10.1097/00005650-199909000-00012

Cao, J. Q., Surgeoner, B., Manna, M., Boileau, J. F., Gelmon, K. A., Brackstone, M., Brezden-Masley, C., Jerzak, K. J., Prakash, I., Sehdev, S., Wong, S. M., Bouganim, N., Cescon, D. W., Chia, S., Dayes, I. S., Joy, A. A., & Henning, J. W. (2024). Guidance for Canadian breast cancer practice: National consensus recommendations for clinical staging of patients newly diagnosed with breast cancer. *Current Oncology, 31*(11), 7226–7243. https://doi.org/10.3390/curroncol31110533

Centre for Epidemiology and Evidence. (2017). *Developing and using program logic: A guide*. Ministry of Health.

Chalmers, K. J., Jorm, A. F., Kelly, C. M., Reavley, N. J., Bond, K. S., Cottrill, F. A., & Wright, J. (2020). Offering mental health first aid to a person after a potentially traumatic event: A Delphi study to redevelop the 2008 guidelines. *BMC Psychology, 8*(1), 105. https://doi.org/10.1186/s40359-020-00473-7

Chan, J., Searle, E. F., Khodyakov, D., Denson, L., Echeverri, A., Browne, E. M., Chiarelli, Y., Dickey, L. L., Erickson, D. S., Flannery, J., Kaplan, L. J., Markovitz, S., Popescu, S. V., & Shenoy, E. S. (2024). They will come, you must build it: A modified Delphi process applied to preparing acute care facilities infrastructure for high-consequence infectious diseases. *Health Security, 22*(5), 384–393. https://doi.org/10.1089/hs.2024.0013

Charles, K. R., Hall, L., Ullman, A. J., & Schults, J. A. (2022). Methodology minute: Utilizing the RAND/UCLA appropriateness method to develop guidelines for infection prevention. *American Journal of Infection Control, 50*(3), 345–348. https://doi.org/10.1016/j.ajic.2021.12.012

Choi, W. S., Sung, Y., Kim, J., Seok, H., Choe, Y. J., Cheong, C., Cho, J., Lee, D. W., Shin, J. Y., & Yu, S. Y. (2024). Prioritization of vaccines for introduction in the national immunization program in the Republic of Korea. *Vaccine, 12*(8), Article 886. https://doi.org/10.3390/vaccines12080886

Clement, S., Jarrett, M., Henderson, C., & Thornicroft, G. (2010). Messages to use in population-level campaigns to reduce mental health-related stigma: Consensus development study. *Epidemiologia e Psichiatria Sociale, 19*(1), 72–79. https://doi.org/10.1017/s1121189x00001627

Cook, J., Oreskes, N., Doran, P. T., Anderegg, W. R. L., Verheggen, B., Maibach, E. W., Carlton, J. S., Lewandowsky, S., Skuce, A. G., Green, S. A., Nuccitelli, D., Jacobs, P., Richardson, M., Winkler, B., Painting, R., & Rice, K. (2016). Consensus on consensus: A synthesis of consensus estimates on human-caused global warming. *Environmental Research Letters, 11*(4), 048002. https://doi.org/10.1088/1748-9326/11/4/048002

Cormack, C. J., Childs, J., & Kent, F. (2024). Competencies required by sonographers teaching ultrasound interprofessionally: A Delphi consensus study. *BMC Medical Education, 24*(1), Article 970. https://doi.org/10.1186/s12909-024-05933-x

Dale, E., Conigrave, K. M., Kelly, P. J., Ivers, R., Clapham, K., & Lee, K. S. K. (2021). A Delphi yarn: Applying Indigenous knowledges to enhance the cultural utility of SMART Recovery Australia [Article]. *Addiction Science & Clinical Practice*, *16*(1), Article 2. https://doi.org/10.1186/s13722-020-00212-8

Dalkey, N. (1969). *The Delphi method: An experimental study of group opinion.* RAND.

Dalkey, N., & Brown, B. (1971). *Comparison of group judgment techniques with short-range predictions and almanac questions.* RAND.

Dalkey, N., & Helmer, O. (1962). *An experimental application of the Delphi method to the use of experts.*

Dalkey, N., & Helmer, O. (1963). An experimental application of the DELPHI method to the use of experts. *Management Science*, *9*(3), 458–467. https://doi.org/10.1287/mnsc.9.3.458

Daniel, C., & Hernandez, T. (2024). What retail apocalypse? A Delphi forecast of commercial space demand in the Toronto region. *Journal of Retailing and Consumer Services*, *77*, 103670. https://doi.org/10.1016/j.jretconser.2023.103670

Daudén, E., Belinchón, I., Colominas-González, E., Coto, P., de la Cueva, P., Gallardo, F., Poveda, J. L., Ramírez, E., Ros, S., Ruíz-Villaverde, R., Comellas, M., & Lizán, L. (2024). Defining well-being in psoriasis: A Delphi consensus among healthcare professionals and patients. *Scientific Reports*, *14*(1), Article 14519. https://doi.org/10.1038/s41598-024-64738-6

Davis-Stober, C. P., Budescu, D. V., Dana, J., & Broomell, S. B. (2014). When is a crowd wise? *Decision*, *1*, 79–101. https://doi.org/10.1037/dec0000004

Dayé, C. (2018). How to train your oracle: The Delphi method and its turbulent youth in operations research and the policy sciences. *Social Studies of Science*, *48*(6), 846–868. https://doi.org/10.1177/0306312718798497

de Oliveira, S., & Nisbett, R. E. (2018). Demographically diverse crowds are typically not much wiser than homogeneous crowds. *Proceedings of the National Academy of Sciences USA*, *115*(9), 2066–2071. https://doi.org/10.1073/pnas.1717632115

Del Grande, C., & Kaczorowski, J. (2023). Rating versus ranking in a Delphi survey: A randomized controlled trial. *Trials*, *24*(1), 543. https://doi.org/10.1186/s13063-023-07442-6

Dezecache, G., Dockendorff, M., Ferreiro, D. N., Deroy, O., & Bahrami, B. (2022). Democratic forecast: Small groups predict the future better than individuals and crowds. *Journal of Experimental Psychology: Applied*, *28*(3), 525–537. https://doi.org/10.1037/xap0000424

Diamond, I. R., Grant, R. C., Feldman, B. M., Pencharz, P. B., Ling, S. C., Moore, A. M., & Wales, P. W. (2014). Defining consensus: A systematic review recommends methodologic criteria for reporting of Delphi studies. *Journal of*

Clinical Epidemiology, *67*(4), 401–409. https://doi.org/10.1016/j. jclinepi.2013.12.002

Dreber, A., Pfeiffer, T., Almenberg, J., Isaksson, S., Wilson, B., Chen, Y., Nosek, B. A., & Johannesson, M. (2015). Using prediction markets to estimate the reproducibility of scientific research. *PNAS*, *112*, 15343–15347. https://doi. org/10.1073/pnas.1516179112

Drennan, J. (2003). Cognitive interviewing: Verbal data in the design and pretesting of questionnaires. *Journal of Advanced Nursing*, *42*(1), 57–63. https:// doi.org/10.1046/j.1365-2648.2003.02579.x

Dubbeldeman, E. M., van der Kleij, R. M. J. J., Brakema, E. A., & Crone, M. R. (2024). Expert consensus on multilevel implementation hypotheses to promote the uptake of youth care guidelines: A Delphi study. *Health Research Policy and Systems*, *22*(1), 89. https://doi.org/10.1186/s12961-024-01167-x

Duijzer, R., Bernts, L. H. P., Geerts, A., van Hoek, B., Coenraad, M. J., Rovers, C., Alvaro, D., Kuijper, E. J., Nevens, F., Halbritter, J., Colmenero, J., Kupcinskas, J., Salih, M., Hogan, M. C., Ronot, M., Vilgrain, V., Hanemaaijer, N. M., Kamath, P. S., Strnad, P., et al. (2024). Clinical management of liver cyst infections: An international, modified Delphi-based clinical decision framework. *The Lancet Gastroenterology & Hepatology*, *9*(9), 884–894. https://doi. org/10.1016/S2468-1253(24)00094-3

Eccles, M. P., & Mittman, B. S. (2006). Welcome to Implementation Science. *Implementation Science*, *1*(1), 1. https://doi.org/10.1186/1748-5908-1-1

Elsman, E. B. M., Mokkink, L. B., Terwee, C. B., Beaton, D., Gagnier, J. J., Tricco, A. C., Baba, A., Butcher, N. J., Smith, M., Hofstetter, C., Aiyegbusi, O. L., Berardi, A., Farmer, J., Haywood, K. L., Krause, K. R., Markham, S., Mayo-Wilson, E., Mehdipour, A., Ricketts, J., et al. (2024). Guideline for reporting systematic reviews of outcome measurement instruments (OMIs): PRISMA-COSMIN for OMIs 2024. *Journal of Patient-Reported Outcomes*, *8*(1), Article 64. https://doi.org/10.1186/s41687-024-00727-7

Fan, T., Zhu, S., Wang, H., Dong, Y., Zhou, Y., Song, Y., Pan, S., Wu, Q., Smith, G. D., Li, Y., & Han, Y. (2024). Development and validation of the self-report symptom inventory of immune-related adverse events in patients with lung cancer. *Asia-Pacific Journal of Oncology Nursing*, *11*(12), 100603. https://doi. org/10.1016/j.apjon.2024.100603

Farina, N., Rajagopalan, J., Alladi, S., Ibnidris, A., Ferri, C. P., Knapp, M., & Comas-Herrera, A. (2024). Estimating the number of people living with dementia at different stages of the condition in India: A Delphi process. *Dementia*, *23*(3), 438–451. https://doi.org/10.1177/14713012231181627

Ferguson, J. H. (1996). The NIH Consensus Development Program. The evolution of guidelines. *International Journal of Technology Assessment in Health Care*, *12*(3), 460–474. https://www.ncbi.nlm.nih.gov/pubmed/8840666

Ferri, C. P., Prince, M., Brayne, C., Brodaty, H., Fratiglioni, L., Ganguli, M., Hall, K., Hasegawa, K., Hendrie, H., Huang, Y., Jorm, A., Mathers, C., Menezes, P. R., Rimmer, E., & Scazufca, M. (2005). Global prevalence of dementia: A Delphi consensus study. *Lancet, 366*(9503), 2112–2117. https://doi. org/10.1016/s0140-6736(05)67889-0

Fischer, J. A., Kelly, C. M., Kitchener, B. A., & Jorm, A. F. (2013). Development of guidelines for adults on how to communicate with adolescents about mental health problems and other sensitive topics: A Delphi study. *SAGE Open, 3*(4), 2158244013516769. https://doi.org/10.1177/2158244013516769

Fitch, K., Bernstein, S. J., Aguilar, M. D., Burnand, B., LaCalle, J. R., & Lazaro, P. (2001). *The RAND/UCLA appropriateness method user's manual.* RAND. https://search.library.wisc.edu/catalog/999911619702121

Foth, T., Efstathiou, N., Vanderspank-Wright, B., Ufholz, L. A., Dütthorn, N., Zimansky, M., & Humphrey-Murto, S. (2016). The use of Delphi and Nominal Group Technique in nursing education: A review. *International Journal of Nursing Studies, 60*, 112–120. https://doi.org/10.1016/j. ijnurstu.2016.04.015

Frank, H. E., Albanese, A., Sun, S., Saadeh, F., Johnson, B. T., Elwy, A. R., & Loucks, E. B. (2024). Mindfulness-based stress reduction health insurance coverage: If, how, and when? An integrated knowledge translation (iKT) Delphi key informant analysis. *Mindfulness, 15*(5), 1220–1233. https://doi. org/10.1007/s12671-024-02366-x

French, B., Babbage, C., Cassidy, S., & Rennick-Egglestone, S. Misrepresentation by online study participants – A threat to data integrity. *Lancet Psychiatry.* https://doi.org/10.1016/S2215-0366(24)00359-6

Frey, V., & Van de Rijt, A. (2021). Social influence undermines the wisdom of crowds in sequential decision making. *Management Science, 67*, 4273–4286. https://doi.org/10.1287/mnsc.2020.3713

Gallegos, D., Booth, S., Pollard, C. M., Chilton, M., & Kleve, S. (2023). Food security definition, measures and advocacy priorities in high-income countries: A Delphi consensus study. *Public Health Nutrition, 26*(10), 1986–1996. https://doi.org/10.1017/s1368980023000915

Galton, F. (1907). Vox populi. *Nature, 75*(1949), 450–451. https://doi. org/10.1038/075450a0

Gargon, E., Crew, R., Burnside, G., & Williamson, P. R. (2019). Higher number of items associated with significantly lower response rates in COS Delphi surveys. *Journal of Clinical Epidemiology, 108*, 110–120. https://doi. org/10.1016/j.jclinepi.2018.12.010

Gattrell, W. T., Logullo, P., van Zuuren, E. J., Price, A., Hughes, E. L., Blazey, P., Winchester, C. C., Tovey, D., Goldman, K., Hungin, A. P., & Harrison, N. (2024). ACCORD (ACcurate COnsensus Reporting Document): A reporting guideline for consensus methods in biomedicine developed via a modified

Delphi. *PLoS Medicine, 21*(1), e1004326. https://doi.org/10.1371/journal. pmed.1004326

Geyle, H. M., Tingley, R., Amey, A. P., Cogger, H., Couper, P. J., Cowan, M., Craig, M. D., Doughty, P., Driscoll, D. A., Ellis, R. J., Emery, J. P., Fenner, A., Gardner, M. G., Garnett, S. T., Gillespie, G. R., Greenlees, M. J., Hoskin, C. J., Keogh, J. S., Lloyd, R., et al. (2021). Reptiles on the brink: Identifying the Australian terrestrial snake and lizard species most at risk of extinction. *Pacific Conservation Biology, 27*(1), 3–12. https://doi.org/10.1071/PC20033

Gibbins, A. K., Wood, P. J., & Spark, M. J. (2017). Managing inappropriate use of non-prescription combination analgesics containing codeine: A modified Delphi study. *Research in Social & Administrative Pharmacy, 13*(2), 369–377. https://doi.org/10.1016/j.sapharm.2016.02.015

Gnatzy, T., Warth, J., von der Gracht, H. A., & Darkow, I. (2011). Validating an innovative real-time Delphi approach – A methodological comparison between real-time and conventional Delphi studies. *Technological Forecasting and Social Change, 78*(9), 1681–1694. https://doi.org/10.1016/j. techfore.2011.04.006

Gordon, T. J., & Helmer, O. (1964). *Report on a long-range forecasting study*. RAND.

Gordon, T., & Pease, A. (2006). RT Delphi: An efficient, "round-less" almost real time Delphi method. *Technological Forecasting and Social Change, 73*(4), 321–333. https://doi.org/10.1016/j.techfore.2005.09.005

Graefe, A., & Armstrong, J. S. (2011). Comparing face-to-face meetings, nominal groups, Delphi and prediction markets on an estimation task. *International Journal of Forecasting, 27*(1), 183–195. https://doi.org/10.1016/j. ijforecast.2010.05.004

Granovskiy, B., Gold, J. M., Sumpter, D. J., & Goldstone, R. L. (2015). Integration of social information by human groups. *Topics in Cognitive Science, 7*(3), 469–493. https://doi.org/10.1111/tops.12150

Grant, S., Booth, M., & Khodyakov, D. (2018). Lack of preregistered analysis plans allows unacceptable data mining for and selective reporting of consensus in Delphi studies. *Journal of Clinical Epidemiology, 99*, 96–105. https://doi. org/10.1016/j.jclinepi.2018.03.007

Gürçay, B., Mellers, B. A., & Baron, J. (2015). The power of social influence on estimation accuracy. *Journal of Behavioral Decision Making, 28*, 250–261. https://doi.org/10.1002/bdm.1843

Harb, S. I., Tao, L., Peláez, S., Boruff, J., Rice, D. B., & Shrier, I. (2021). Methodological options of the nominal group technique for survey item elicitation in health research: A scoping review. *Journal of Clinical Epidemiology, 139*, 140–148. https://doi.org/10.1016/j.jclinepi.2021.08.008

Harris, R. (2021, June 29). Climate explained: How the IPCC reaches scientific consensus on climate change. *The Conversation.* https://theconversation.

com/climate-explained-how-the-ipcc-reaches-scientific-consensus-on-climate-change-162600

Hart, L. M., Jorm, A. F., Kanowski, L. G., Kelly, C. M., & Langlands, R. L. (2009). Mental health first aid for Indigenous Australians: Using Delphi consensus studies to develop guidelines for culturally appropriate responses to mental health problems. *BMC Psychiatry, 9,* 47. https://doi.org/10.118 6/1471-244x-9-47

Hart, L. M., Mason, R. J., Kelly, C. M., Cvetkovski, S., & Jorm, A. F. (2016). 'teen Mental Health First Aid': A description of the program and an initial evaluation. *International Journal of Mental Health Systems, 10,* 3. https://doi.org/10.1186/s13033-016-0034-1

Hart, L. M., Morgan, A. J., Rossetto, A., Kelly, C. M., Gregg, K., Gross, M., Johnson, C., & Jorm, A. F. (2022). teen Mental Health First Aid: 12-month outcomes from a cluster crossover randomized controlled trial evaluation of a universal program to help adolescents better support peers with a mental health problem. *BMC Public Health, 22*(1), 1159. https://doi.org/10.1186/ s12889-022-13554-6

Hartstein, L. E., Mathew, G. M., Reichenberger, D. A., Rodriguez, I., Allen, N., Chang, A. M., Chaput, J. P., Christakis, D. A., Garrison, M., Gooley, J. J., Koos, J. A., Van Den Bulck, J., Woods, H., Zeitzer, J. M., Dzierzewski, J. M., & Hale, L. (2024). The impact of screen use on sleep health across the lifespan: A National Sleep Foundation consensus statement. *Sleep Health, 10*(4), 373–384. https://doi.org/10.1016/j.sleh.2024.05.001

Hasson, F., Keeney, S., & McKenna, H. (2000). Research guidelines for the Delphi survey technique. *Journal of Advanced Nursing, 32*(4), 1008–1015.

Healy, D. (2006). Manufacturing consensus. *Culture, Medicine and Psychiatry, 30*(2), 135–156. https://doi.org/10.1007/s11013-006-9013-3

Hemming, V., Burgman, M. A., Hanea, A. M., McBride, M. F., & Wintle, B. C. (2018a). A practical guide to structured expert elicitation using the IDEA protocol. *Methods in Ecology and Evolution, 9*(1), 169–180. https://doi.org/1 0.1111/2041-210X.12857

Hemming, V., Walshe, T. V., Hanea, A. M., Fidler, F., & Burgman, M. A. (2018b). Eliciting improved quantitative judgements using the IDEA protocol: A case study in natural resource management. *PLoS One, 13*(6), e0198468. https:// doi.org/10.1371/journal.pone.0198468

Hill, K. Q., & Fowles, J. (1975). The methodological worth of the Delphi fore-casting technique. *Technological Forecasting and Social Change, 7*(2), 179–192. https://doi.org/10.1016/0040-1625(75)90057-8

Humphrey-Murto, S., & de Wit, M. (2019). The Delphi method-more research please. *Journal of Clinical Epidemiology, 106,* 136–139. https://doi.org/10.1016/j.jclinepi.2018.10.011

Humphrey-Murto, S., Varpio, L., Wood, T. J., Gonsalves, C., Ufholz, L. A., Mascioli, K., Wang, C., & Foth, T. (2017). The use of the Delphi and other consensus goup methods in medical education research: A review. *Academic Medicine*, *92*(10), 1491–1498. https://doi.org/10.1097/acm.0000000000001812

ISO. (2023). *ISO/TC 48 laboratory equipment*. International Organization for Standardization. Retrieved February 19, 2023, from https://www.iso.org/committee/48908.html

Janis, I. L. (1972). *Victims of groupthink*. Houghton Mifflin.

Jorm, A. F. (2015). Using the Delphi expert consensus method in mental health research. *The Australian and New Zealand Journal of Psychiatry*, *49*(10), 887–897. https://doi.org/10.1177/0004867415600891

Jorm, A. (2025). *Expert consensus in science*. Palgrave Macmillan. https://doi.org/10.1007/978-981-97-9222-1

Jorm, A. F., Kitchener, B. A., & Reavley, N. J. (2019). Mental Health First Aid training: Lessons learned from the global spread of a community education program. *World Psychiatry*, *18*(2), 142–143. https://doi.org/10.1002/wps.20621

Jorm, A. F., & Ross, A. M. (2018). Guidelines for the public on how to provide mental health first aid: Narrative review. *BJPsych Open*, *4*(6), 427–440. https://doi.org/10.1192/bjo.2018.58

Jünger, S., Payne, S. A., Brine, J., Radbruch, L., & Brearley, S. G. (2017). Guidance on Conducting and REporting DElphi Studies (CREDES) in palliative care: Recommendations based on a methodological systematic review. *Palliative Medicine*, *31*(8), 684–706. https://doi.org/10.1177/0269216317690685

Kaufmann, A., & Gupta, M. M. (1988). *Fuzzy mathematical models in engineering and management science*. Elsevier.

Kearney, A., Williamson, P., Young, B., Bagley, H., Gamble, C., Denegri, S., Muir, D., Simon, N. A., Thomas, S., Elliot, J. T., Bulbeck, H., Crocker, J. C., Planner, C., Vale, C., Clarke, M., Sprosen, T., & Woolfall, K. (2017). Priorities for methodological research on patient and public involvement in clinical trials: A modified Delphi process. *Health Expectations*, *20*(6), 1401–1410. https://doi.org/10.1111/hex.12583

Keck, S., & Tang, W. (2020). Enhancing the wisdom of the crowd with cognitive-process diversity: The benefits of aggregating intuitive and analytical judgments. *Psychological Science*, *31*(10), 1272–1282. https://doi.org/10.1177/0956797620941840

Kepp, K. P., Aavitsland, P., Ballin, M., Balloux, F., Baral, S., Bardosh, K., Bauchner, H., Bendavid, E., Bhopal, R., Blumstein, D. T., Boffetta, P., Bourgeois, F., Brufsky, A., Collignon, P. J., Cripps, S., Cristea, I. A., Curtis, N., Djulbegovic, B., Faude, O., et al. (2024). Panel stacking is a threat to consensus statement validity. *Journal of Clinical Epidemiology*, *173*, Article 111428. https://doi.org/10.1016/j.jclinepi.2024.111428

Khodyakov, D., Grant, S., Kroger, J., & Bauman, M. (2023a). *RAND methodological guidance for conducting and critically appraising Delphi panels*. RAND Corporation.

Khodyakov, D., Grant, S., Kroger, J., Gadwah-Meaden, C., Motala, A., & Larkin, J. (2023b). Disciplinary trends in the use of the Delphi method: A bibliometric analysis. *PLoS One, 18*(8), e0289009. https://doi.org/10.1371/journal.pone.0289009

Kitchener, B. A., & Jorm, A. F. (2002). *Mental health first aid manual*. Centre for Mental Health Research.

Kitchener, B. A., Jorm, A. F., & Kelly, C. M. (2010). *Mental health first aid manual* (2nd ed.). Orygen Youth Health Research Centre.

Koppold, D. A., Breinlinger, C., Hanslian, E., Kessler, C., Cramer, H., Khokhar, A. R., Peterson, C. M., Tinsley, G., Vernieri, C., Bloomer, R. J., Boschmann, M., Bragazzi, N. L., Brandhorst, S., Gabel, K., Goldhamer, A. C., Grajower, M. M., Harvie, M., Heilbronn, L., Horne, B. D., et al. (2024). International consensus on fasting terminology. *Cell Metabolism, 36*(8), 1779–1794.e1774. https://doi.org/10.1016/j.cmet.2024.06.013

Kuusi, O. (1999). *Expertise in the future use of generic technologies: Epistemic and methodological considerations concerning Delphi studies*. Government Institute for Economic Research.

Lakens, D., Mesquida, C., Rasti, S., & Ditroilo, M. (2024). The benefits of pre-registration and Registered Reports. *Evidence-Based Toxicology, 2*(1), 2376046. https://doi.org/10.1080/2833373X.2024.2376046

Lambert, S. D., Ould Brahim, L., Morrison, M., Girgis, A., Yaffe, M., Belzile, E., Clayberg, K., Robinson, J., Thorne, S., Bottorff, J. L., Duggleby, W., Campbell-Enns, H., Kim, Y., & Loiselle, C. G. (2019). Priorities for caregiver research in cancer care: An international Delphi survey of caregivers, clinicians, managers, and researchers. *Supportive Care in Cancer, 27*(3), 805–817. https://doi.org/10.1007/s00520-018-4314-y

Lancet. (2022). *Information for authors*. www.thelancet.com

Landeta, J., & Lertxundi, A. (2024). Quality indicators for Delphi studies. *FUTURES & FORESIGHT SCIENCE, 6*(1), e172. https://doi.org/10.1002/ffo2.172

Langley, T., Young, E., Hunter, A., & Bains, M. (2024). Developing a vape shop-based smoking cessation intervention: A Delphi study. *Nicotine and Tobacco Research, 26*(10), 1362–1369. https://doi.org/10.1093/ntr/ntae105

Lazarus, J. V., Romero, D., Kopka, C. J., Karim, S. A., Abu-Raddad, L. J., Almeida, G., Baptista-Leite, R., Barocas, J. A., Barreto, M. L., Bar-Yam, Y., Bassat, Q., Batista, C., Bazilian, M., Chiou, S. T., Del Rio, C., Dore, G. J., Gao, G. F., Gostin, L. O., Hellard, M., et al. (2022). A multinational Delphi consensus to end the COVID-19 public health threat. *Nature, 611*(7935), 332–345. https://doi.org/10.1038/s41586-022-05398-2

Lecours, A. (2020). Scientific, professional and experiential validation of the model of preventive behaviours at work: Protocol of a modified Delphi Study. *BMJ Open, 10*(9), e035606. https://doi.org/10.1136/bmjopen-2019-035606

Lee, C., Song, H., & Mjelde, J. W. (2008). The forecasting of International Expo tourism using quantitative and qualitative techniques. *Tourism Management, 29*(6), 1084–1098. https://doi.org/10.1016/j.tourman.2008.02.007

Lei, F. (2024). Online recruitment for an online survey study: Our experience of dealing with fraudsters. *Applied Nursing Research, 80*, 151854. https://doi.org/10.1016/j.apnr.2024.151854

Li, E. Y., Tung, C. Y., & Chang, S. H. (2016). The wisdom of crowds in action: Forecasting epidemic diseases with a web-based prediction market system. *International Journal of Medical Informatics, 92*, 35–43. https://doi.org/10.1016/j.ijmedinf.2016.04.014

Linares, O., Martínez-Jauregui, M., Carranza, J., & Soliño, M. (2024). Bridging sustainable game management into land use policy: From principles to practice [Article]. *Land Use Policy, 145*, Article 107269. https://doi.org/10.1016/j.landusepol.2024.107269

Linstone, H. A., & Turoff, M. (2002). *The Delphi method: Techniques and applications.* Addison-Wesley Publishing Company, Advanced Book Program. https://books.google.com.au/books?id=uZ0RkAEACAAJ

Liu, X., Cruz Rivera, S., Moher, D., Calvert, M. J., Denniston, A. K., Spirit, A. I., & Group, C.-A. W. (2020). Reporting guidelines for clinical trial reports for interventions involving artificial intelligence: The CONSORT-AI extension. *Lancet Digital Health, 2*(10), e537–e548. https://doi.org/10.1016/S2589-7500(20)30218-1

Locke, C. C., & Anderson, C. (2015). The downside of looking like a leader: Power, nonverbal confidence, and participative decision-making. *Journal of Experimental Social Psychology, 58*, 42–47. https://doi.org/10.1016/j.jesp.2014.12.004

Lu, M. S., Cheng, C. C., Lin, C. F., Yang, C. M., & Liao, M. N. (2024). Revising the Taiwan code of ethics for nurses. *The Journal of Nursing, 71*(4), 32–43. https://doi.org/10.6224/JN.202408_71(4).06

MacLennan, S., Kirkham, J., Lam, T. B. L., & Williamson, P. R. (2018). A randomized trial comparing three Delphi feedback strategies found no evidence of a difference in a setting with high initial agreement. *Journal of Clinical Epidemiology, 93*, 1–8. https://doi.org/10.1016/j.jclinepi.2017.09.024

Mannes, A. E., Soll, J. B., & Larrick, R. P. (2014). The wisdom of select crowds. *Journal of Personality and Social Psychology, 107*, 276–299. https://doi.org/10.1037/a0036677

Manyara, A. M., Purvis, A., Ciani, O., Collins, G. S., & Taylor, R. S. (2024). Sample size in multistakeholder Delphi surveys: At what minimum sample size do replicability of results stabilize? *Journal of Clinical Epidemiology, 174,* Article 111485. https://doi.org/10.1016/j.jclinepi.2024.111485

Martel, C., Allen, J., Pennycook, G., & Rand, D. G. (2024). Crowds can effectively identify misinformation at scale. *Perspectives in Psychological Science, 19,* 477–488. https://doi.org/10.1177/17456916231190388

Mazariego, C., Jefford, M., Chan, R. J., Roberts, N., Millar, L., Anazodo, A., Hayes, S., Brown, B., Saunders, C., Webber, K., Vardy, J., Girgis, A., & Koczwara, B. (2022). Priority recommendations for the implementation of patient-reported outcomes in clinical cancer care: A Delphi study. *Journal of Cancer Survivorship, 16*(1), 33–43. https://doi.org/10.1007/s11764-021-01135-2

McAdam, M. K., Baldessarini, R. J., Murphy, A. L., & Gardner, D. M. (2023). Second international consensus study of antipsychotic dosing (ICSAD-2). *Journal of Psychopharmacology, 37*(10), 982–991. https://doi.org/10.1177/02698811231205688

Meijering, J. V., & Tobi, H. (2016). The effect of controlled opinion feedback on Delphi features: Mixed messages from a real-world Delphi experiment. *Technological Forecasting and Social Change, 103,* 166–173. https://doi.org/10.1016/j.techfore.2015.11.008

Meijering, J. V., & Tobi, H. (2018). The effects of feeding back experts' own initial ratings in Delphi studies: A randomized trial. *International Journal of Forecasting, 34*(2), 216–224. https://doi.org/10.1016/j.ijforecast.2017.11.010

Mellers, B., Ungar, L., Baron, J., Ramos, J., Gurcay, B., Fincher, K., Scott, S. E., Moore, D., Atanasov, P., Swift, S. A., Murray, T., Stone, E., & Tetlock, P. E. (2014). Psychological strategies for winning a geopolitical forecasting tournament. *Psychological Science, 25*(5), 1106–1115. https://doi.org/10.1177/0956797614524255

Mercier, H., & Claidière, N. (2022). Does discussion make crowds any wiser? *Cognition, 222,* 104912. https://doi.org/10.1016/j.cognition.2021.104912

Metry, D., Copp, H. L., Rialon, K. L., Iacobas, I., Baselga, E., Dobyns, W. B., Drolet, B., Frieden, I. J., Garzon, M., Haggstrom, A., Hanson, D., Hollenbach, L., Keppler-Noreuil, K. M., Maheshwari, M., Siegel, D. H., Waseem, S., & Dias, M. (2024). Delphi consensus on diagnostic criteria for LUMBAR syndrome. *Journal of Pediatrics, 272,* 114101. https://doi.org/10.1016/j.jpeds.2024.114101

Minas, H., & Jorm, A. F. (2010). Where there is no evidence: Use of expert consensus methods to fill the evidence gap in low-income countries and cultural minorities. *International Journal of Mental Health Systems, 4,* 33. https://doi.org/10.1186/1752-4458-4-33

Moodley, S. V., Wolvaardt, J., & Grobler, C. (2024). Developing mental health curricula and a service provision model for clinical associates in South Africa: A

Delphi survey of family physicians and psychiatrists. *BMC Medical Education*, *24*(1), Article 669. https://doi.org/10.1186/s12909-024-05637-2

Morgan, T. L., Fortier, M. S., Jain, R., Lane, K. N., Maclaren, K., McFadden, T., Prorok, J., Robison, J., Weston, Z. J., & Tomasone, J. R. (2025). Development of the Whole Day Matters Toolkit for Primary Care: A consensus-building study to mobilize national public health guidelines in practice. *Health Promotion and Chronic Disease Prevention in Canada*, *45*(1), 1–19. https://doi.org/10.24095/hpcdp.45.1.01

Muchnik, L., Aral, S., & Taylor, S. J. (2013). Social influence bias: A randomized experiment. *Science*, *341*, 647–651. https://doi.org/10.1126/science.1240466

National Institutes of Health. (2023). *NIH consensus development program.* Retrieved February 21, 2023, from https://consensus.nih.gov/

National Public Health Partnership. (2000). *Public health practice in Australia today: A statement of core functions.* National Public Health Partnership.

Navajas, J., Niella, T., Garbulsky, G., Bahrami, B., & Sigman, M. (2018). Aggregated knowledge from a small number of debates outperforms the wisdom of large crowds. *Nature Human Behaviour*, *2*, 126–132. https://doi.org/10.1038/s41562-017-0273-4

Needham, D. M., Sepulveda, K. A., Dinglas, V. D., Chessare, C. M., Friedman, L. A., Bingham, C. O., III, & Turnbull, A. E. (2017). Core outcome measures for clinical research in acute respiratory failure survivors. An international modified Delphi consensus study. *American Journal of Respiratory and Critical Care Medicine*, *196*(9), 1122–1130. https://doi.org/10.1164/rccm.201702-0372OC

Niederberger, M., & Homberg, A. (2023). Argument-based QUalitative Analysis strategy (AQUA) for analyzing free-text responses in health sciences Delphi studies. *MethodsX, 10*, 102156. https://doi.org/10.1016/j.mex.2023.102156

Niederberger, M., Schifano, J., Deckert, S., Hirt, J., Homberg, A., Köberich, S., Kuhn, R., Rommel, A., Sonnberger, M., Alam, M., Backman, C., Banno, M., Bartoszko, J., Bloomfield, F., Bober, M. B., Chalkoo, M., Chan, T. M., Chen, Y., Coscia, C., et al. (2024). Delphi studies in social and health sciences— Recommendations for an interdisciplinary standardized reporting (DELPHISTAR). Results of a Delphi study. *PLoS One, 19*(8), Article e0304651. https://doi.org/10.1371/journal.pone.0304651

Niederberger, M., & Spranger, J. (2020). Delphi technique in health sciences: A map. *Frontiers in Public Health*, *8*, 457. https://doi.org/10.3389/fpubh.2020.00457

Nilsson, K., Andersson, G., Johansson, P., & Lundgren, J. (2023). Developing and designing an internet-based support and education program for patients awaiting kidney transplantation with deceased donors through: A Delphi study. *BMC Nephrology, 24*(1), 311. https://doi.org/10.1186/s12882-023-03364-2

Nosek, B. A., Ebersole, C. R., DeHaven, A. C., & Mellor, D. T. (2018). The pre-registration revolution. *Proceedings of the National Academy of Sciences USA*, *115*(11), 2600–2606. https://doi.org/10.1073/pnas.1708274114

O'Leary, D. E. (2017). Crowd performance in prediction of the World Cup 2014. *European Journal of Operational Research*, *260*, 715–724. https://doi.org/10.1016/j.ejor.2016.12.043

Page, S. E. (2007). *The difference: How the power of diversity creates better groups, firms, schools, and societies.* Princeton University Press.

Page, M. J., McKenzie, J. E., Bossuyt, P. M., Boutron, I., Hoffmann, T., Mulrow, C. D., Shamseer, L., & Moher, D. (2018). *Updating the PRISMA reporting guideline for systematic reviews and meta-analyses: Study protocol.* Monash University School of Public Health and Preventive Medicine.

Page, M. J., McKenzie, J. E., Bossuyt, P. M., Boutron, I., Hoffmann, T. C., Mulrow, C. D., Shamseer, L., Tetzlaff, J. M., Akl, E. A., Brennan, S. E., Chou, R., Glanville, J., Grimshaw, J. M., Hrobjartsson, A., Lalu, M. M., Li, T., Loder, E. W., Mayo-Wilson, E., McDonald, S., et al. (2021). The PRISMA 2020 statement: An updated guideline for reporting systematic reviews. *BMJ*, *372*, n71. https://doi.org/10.1136/bmj.n71

Päivärinta, T., Pekkola, S., & Moe, C. (2011). Grounding theory from Delphi studies. In *International conference on information systems 2011, ICIS 2011*.

Parks, S., d'Angelo, C., & Gunashekar, S. (2018). *Citizen science: Generating ideas and exploring consensus.* The Healthcare Improvement Studies Institute.

Paul, C. L., Sanson-Fisher, R., Douglas, H. E., Clinton-McHarg, T., Williamson, A., & Barker, D. (2011). Cutting the research pie: A value-weighting approach to explore perceptions about psychosocial research priorities for adults with haematological cancers. *European Journal of Cancer Care*, *20*(3), 345–353. https://doi.org/10.1111/j.1365-2354.2010.01188.x

Pfeiffer, T., & Almenberg, J. (2010). Prediction markets and their potential role in biomedical research – A review. *Biosystems*, *102*(2–3), 71–76. https://doi.org/10.1016/j.biosystems.2010.09.005

Prokesch, T., von der Gracht, H. A., & Wohlenberg, H. (2015). Integrating prediction market and Delphi methodology into a foresight support system—Insights from an online game. *Technological Forecasting and Social Change*, *97*, 47–64. https://doi.org/10.1016/j.techfore.2014.02.021

Qin, X., Gao, X., Yang, Y., Ou, S., Luo, J., Wei, H., & Jiang, Q. (2024). Developing a risk assessment tool for cancer-related venous thrombosis in China: A modified Delphi-analytic hierarchy process study. *BMC Cancer*, *24*(1), Article 120. https://doi.org/10.1186/s12885-024-11877-8

Quirke, F. A., Battin, M. R., Bernard, C., Biesty, L., Bloomfield, F. H., Daly, M., Finucane, E., Haas, D. M., Healy, P., Hurley, T., Koskei, S., Meher, S., Molloy, E. J., Niaz, M., Bhraonáin, E. N., Okaronon, C. O., Tabassum, F., Walker, K., Webbe, J. R. H., et al. (2023). Multi-Round versus Real-Time Delphi survey approach for achieving consensus in the COHESION core outcome set: A randomised trial. *Trials, 24*(1), 461. https://doi.org/10.1186/s13063-023-07388-9

Radomski, T. R., Decker, A., Khodyakov, D., Thorpe, C. T., Hanlon, J. T., Roberts, M. S., Fine, M. J., & Gellad, W. F. (2022). Development of a metric to detect and decrease low-value prescribing in older adults. *JAMA Network Open, 5*(2), e2148599. https://doi.org/10.1001/jamanetworkopen.2021.48599

Ramke, J., Evans, J. R., Habtamu, E., Mwangi, N., Silva, J. C., Swenor, B. K., Congdon, N., Faal, H. B., Foster, A., Friedman, D. S., Gichuhi, S., Jonas, J. B., Khaw, P. T., Kyari, F., Murthy, G. V. S., Wang, N., Wong, T. Y., Wormald, R., Yusufu, M., et al. (2022). Grand challenges in global eye health: A global prioritisation process using Delphi method. *Lancet Healthy Longevity, 3*(1), e31–e41. https://doi.org/10.1016/s2666-7568(21)00302-0

RAND. (2025). *Our history*. RAND. Retrieved February 8, 2025, from https://www.rand.org/about/history.html#:~:text=RAND%27s%20articles%20of%20incorporation%20commit,contract%20for%20Project%20RAND%2C%201946

Rauch, W. (1979). The decision Delphi. *Technological Forecasting and Social Change, 15*(3), 159–169. https://doi.org/10.1016/0040-1625(79)90011-8

Robinson, J., Thorn, P., McKay, S., Hemming, L., Battersby-Coulter, R., Cooper, C., Veresova, M., Li, A., Reavley, N., Rice, S., Lamblin, M., Pirkis, J., Reidenberg, D., Harrison, V., Skehan, J., & La Sala, L. (2023). #chatsafe 2.0. Updated guidelines to support young people to communicate safely online about self-harm and suicide: A Delphi expert consensus study. *PLoS One, 18*(8), e0289494. https://doi.org/10.1371/journal.pone.0289494

Ross, A. M., Hart, L. M., Jorm, A. F., Kelly, C. M., & Kitchener, B. A. (2012). Development of key messages for adolescents on providing basic mental health first aid to peers: A Delphi consensus study. *Early Intervention in Psychiatry, 6*(3), 229–238. https://doi.org/10.1111/j.1751-7893.2011.00331.x

Ross, A. M., Kelly, C. M., & Jorm, A. F. (2014). Re-development of mental health first aid guidelines for suicidal ideation and behaviour: A Delphi study. *BMC Psychiatry, 14*, 241. https://doi.org/10.1186/s12888-014-0241-8

Rowe, G., & Wright, G. (1996). The impact of task characteristics on the performance of structured group forecasting techniques. *International Journal of Forecasting, 12*(1), 73–89. https://doi.org/10.1016/0169-2070(95)00658-3

Rowe, G., Wright, G., & McColl, A. (2005). Judgment change during Delphi-like procedures: The role of majority influence, expertise, and confidence. *Technological Forecasting and Social Change, 72*(4), 377–399. https://doi.org/10.1016/j.techfore.2004.03.004

Rubin, G., De Wit, N., Meineche-Schmidt, V., Seifert, B., Hall, N., & Hungin, P. (2006). The diagnosis of IBS in primary care: Consensus development using nominal group technique. *Family Practice, 23*(6), 687–692. https://doi.org/10.1093/fampra/cml050

Ryan, J. M., Devane, D., Simiceva, A., Eppich, W., Kavanagh, D. O., Cullen, C., Hogan, A. M., & McNamara, D. A. (2024). Surgical Handover Core Outcome Measures (SH-CORE): A protocol for the development of a core outcome set for trials in surgical handover [Article]. *Trials, 25*(1), Article 373. https://doi.org/10.1186/s13063-024-08201-x

Sackman, H. (1974). *Delphi assessment: Expert opinion, forecasting, and group process.* RAND.

Sackman, H. (1975). Summary evaluation of Delphi. *Policy Analysis, 1*(4), 693–718. http://www.jstor.org/stable/42784280

Sahle, B. W., Reavley, N. J., Morgan, A. J., Yap, M. B. H., Reupert, A., & Jorm, A. F. (2022). A Delphi study to identify intervention priorities to prevent the occurrence and reduce the impact of adverse childhood experiences. *The Australian and New Zealand Journal of Psychiatry, 56*(6), 686–694. https://doi.org/10.1177/00048674211025717

Saunders, H., Anderson, C., Feldman, F., Holroyd-Leduc, J., Jain, R., Liu, B., Macaulay, S., Marr, S., Silvius, J., Weldon, J., Bayoumi, A. M., Straus, S. E., Tricco, A. C., & Isaranuwatchai, W. (2023). Developing a fall prevention intervention economic model. *PLoS One, 18*(1 January), Article e0280572. https://doi.org/10.1371/journal.pone.0280572

Schifano, J., & Niederberger, M. (2025). How Delphi studies in the health sciences find consensus: A scoping review. *Systematic Reviews, 14*(1), 14. https://doi.org/10.1186/s13643-024-02738-3

Shi, F., Teplitskiy, M., Duede, E., & Evans, J. A. (2019). The wisdom of polarized crowds. *Nature Human Behaviour, 3*(4), 329–336. https://doi.org/10.1038/s41562-019-0541-6

Sii, S., Barton, K., Pasquale, L. R., Yamamoto, T., King, A. J., & Azuara-Blanco, A. (2018). Reporting harm in glaucoma surgical trials: Systematic review and a consensus-derived new classification system. *American Journal of Ophthalmology, 194*, 153–162. https://doi.org/10.1016/j.ajo.2018.07.014

Simoiu, C., Sumanth, C., Mysore, A., & Goel, S. (2019). Studying the "wisdom of crowds" at scale. In *Seventh AAAI conference on human computation and crowdsourcing (HCOMP-19).*

Sinha, I. P., Smyth, R. L., & Williamson, P. R. (2011). Using the Delphi technique to determine which outcomes to measure in clinical trials: Recommendations for the future based on a systematic review of existing studies. *PLoS Medicine, 8*(1), e1000393. https://doi.org/10.1371/journal.pmed.1000393

Smelt, A. F., Louter, M. A., Kies, D. A., Blom, J. W., Terwindt, G. M., van der Heijden, G. J., De Gucht, V., Ferrari, M. D., & Assendelft, W. J. (2014). What do patients consider to be the most important outcomes for effectiveness studies on migraine treatment? Results of a Delphi study. *PLoS One, 9*(6), e98933. https://doi.org/10.1371/journal.pone.0098933

Spranger, J., Homberg, A., Sonnberger, M., & Niederberger, M. (2022). Reporting guidelines for Delphi techniques in health sciences: A methodological review. *Zeitschrift für Evidenz, Fortbildung und Qualität im Gesundheitswesen, 172*, 1–11. https://doi.org/10.1016/j.zefq.2022.04.025

Sri, A., Bailey, K. E., Scarborough, R., Gilkerson, J. R., Thursky, K., Browning, G. F., & Hardefeldt, L. Y. (2024). Reaching consensus amongst international experts on the use of high importance-rated antimicrobials in animals – A Delphi study. *One Health, 19*, Article 100883. https://doi.org/10.1016/j.onehlt.2024.100883

State of Queensland. (2017). *2017 consensus statement: Land use impacts on Great Barrier Reef water quality and ecosystem condition*. State of Queensland. https://researchonline.jcu.edu.au/50116/1/2017-scientific-consensus-statement-summary.pdf

Steinert, M. (2009). A dissensus based online Delphi approach: An explorative research tool. *Technological Forecasting and Social Change, 76*(3), 291–300. https://doi.org/10.1016/j.techfore.2008.10.006

Sulik, J., Bahrami, B., & Deroy, O. (2022). The diversity gap: When diversity matters for knowledge. *Perspectives on Psychological Science, 17*(3), 752–767. https://doi.org/10.1177/17456916211006070

Surowiecki, J. (2004). *The wisdom of crowds: Why the many are smarter than the few*. Doubleday.

Swedo, S. E., Baguley, D. M., Denys, D., Dixon, L. J., Erfanian, M., Fioretti, A., Jastreboff, P. J., Kumar, S., Rosenthal, M. Z., Rouw, R., Schiller, D., Simner, J., Storch, E. A., Taylor, S., Werff, K. R. V., Altimus, C. M., & Raver, S. M. (2022). Consensus definition of misophonia: A Delphi study. *Frontiers in Neuroscience, 16*, 841816. https://doi.org/10.3389/fnins.2022.841816

Tapio, P. (2003). Disaggregative policy Delphi: Using cluster analysis as a tool for systematic scenario formation. *Technological Forecasting and Social Change, 70*(1), 83–101. https://doi.org/10.1016/S0040-1625(01)00177-9

Ter Veer, E., van Rijssen, L. B., Besselink, M. G., Mali, R. M. A., Berlin, J. D., Boeck, S., Bonnetain, F., Chau, I., Conroy, T., Van Cutsem, E., Deplanque, G., Friess, H., Glimelius, B., Goldstein, D., Herrmann, R., Labianca, R., Van Laethem, J. L., Macarulla, T., van der Meer, J. H. M., et al. (2018). Consensus statement on mandatory measurements in pancreatic cancer trials (COMM-PACT) for systemic treatment of unresectable disease. *Lancet Oncology, 19*(3), e151–e160. https://doi.org/10.1016/s1470-2045(18)30098-6

Thoomes, E., Falla, D., Cleland, J. A., Fernández-de-Las-Peñas, C., Gallina, A., & de Graaf, M. (2023). Conservative management for lumbar radiculopathy based on the stage of the disorder: A Delphi study. *Disability and Rehabilitation, 45*(21), 3539–3548. https://doi.org/10.1080/09638288.2022.2130448

Toulany, A., Khodyakov, D., Mooney, S., Stromquist, L., Bailey, K., Barber, C. E., Batthish, M., Cleverley, K., Dimitropoulos, G., Gorter, J. W., Grahovac, D.,

Grimes, R., Guttman, B., Hébert, M. L., John, T., Lo, L., Luong, D., MacGregor, L., Mukerji, G., et al. (2024). Quality indicators for transition from pediatric to adult care for youth with chronic conditions: Proposal for an online modified Delphi study. *JMIR Research Protocols, 13*, e60860. https://doi.org/10.2196/60860

Tournier, A. L., Bonamin, L. V., Buchheim-Schmidt, S., Cartwright, S., Dombrowsky, C., Doesburg, P., Holandino, C., Kokornaczyk, M. O., van de Kraats, E. B., López-Carvallo, J. A., Nandy, P., Mazón-Suástegui, J. M., Mirzajani, F., Poitevin, B., Scherr, C., Thieves, K., Würtenberger, S., & Baumgartner, S. (2024). Scientific guidelines for preclinical research on potentised preparations manufactured according to current pharmacopoeias—The PrePoP guidelines. *Journal of Integrative Medicine, 22*(5), 533–544. https://doi.org/10.1016/j.joim.2024.06.005

Toyokawa, W., Whalen, A., & Laland, K. N. (2019). Social learning strategies regulate the wisdom and madness of interactive crowds. *Nature Human Behaviour, 3*(2), 183–193. https://doi.org/10.1038/s41562-018-0518-x

Turnbull, A. E., Dinglas, V. D., Friedman, L. A., Chessare, C. M., Sepúlveda, K. A., Bingham, C. O., III, & Needham, D. M. (2018). A survey of Delphi panelists after core outcome set development revealed positive feedback and methods to facilitate panel member participation. *Journal of Clinical Epidemiology, 102*, 99–106. https://doi.org/10.1016/j.jclinepi.2018.06.007

Turoff, M. (1970). The design of a policy Delphi. *Technological Forecasting and Social Change, 2*(2), 149–171. https://doi.org/10.1016/0040-1625(70)90161-7

Turoff, M. (1971). Delphi conferencing: Computer-based conferencing with anonymity. *Technological Forecasting and Social Change, 3*, 159–204. https://doi.org/10.1016/S0040-1625(71)80012-4

Vaesen, K., Dusseldorp, G. L., & Brandt, M. J. (2021). An emerging consensus in palaeoanthropology: Demography was the main factor responsible for the disappearance of Neanderthals. *Scientific Reports, 11*(1), 4925. https://doi.org/10.1038/s41598-021-84410-7

Waggoner, J., Carline, J. D., & Durning, S. J. (2016). Is there a consensus on consensus methodology? Descriptions and recommendations for future consensus research. *Academic Medicine, 91*(5), 663–668. https://doi.org/10.1097/acm.0000000000001092

Webler, T., Levine, D., Rakel, H., & Renn, O. (1991). A novel approach to reducing uncertainty: The group Delphi. *Technological Forecasting and Social Change, 39*(3), 253–263. https://doi.org/10.1016/0040-1625(91)90040-M

Yap, M. B. H., Fowler, M., Reavley, N., & Jorm, A. F. (2015). Parenting strategies for reducing the risk of childhood depression and anxiety disorders: A Delphi consensus study. *Journal of Affective Disorders, 183*, 330–338. https://doi.org/10.1016/j.jad.2015.05.031

Yap, M. B. H., & Jorm, A. F. (2015). Parental factors associated with childhood anxiety, depression, and internalizing problems: A systematic review and meta-analysis. *Journal of Affective Disorders, 175,* 424–440. https://doi.org/10.1016/j.jad.2015.01.050

Yap, M. B., Pilkington, P. D., Ryan, S. M., Kelly, C. M., & Jorm, A. F. (2014). Parenting strategies for reducing the risk of adolescent depression and anxiety disorders: A Delphi consensus study. *Journal of Affective Disorders, 156,* 67–75. https://doi.org/10.1016/j.jad.2013.11.017

Yu, M., Keel, S., Mariotti, S., Mills, J., & Müller, A. (2023). Development of the WHO eye care competency framework. *Human Resources for Health, 21*(1), 46. https://doi.org/10.1186/s12960-023-00834-4

Zachar, P., & Kendler, K. S. (2012). The removal of Pluto from the class of planets and homosexuality from the class of psychiatric disorders: A comparison. *Philosophy, Ethics, and Humanities in Medicine, 7,* 4. https://doi.org/10.118 6/1747-5341-7-4

Zickafoose, A., Lu, P., & Baker, M. (2022). Forecasting food innovations with a Delphi study. *Food, 11*(22). https://doi.org/10.3390/foods11223723

INDEX

© The Author(s) 2025

A. Jorm, *Using the Delphi Method to Establish Expert Consensus,*

Advancing Methods for Interdisciplinarity in the Social Sciences,

https://doi.org/10.1007/978-981-96-8357-4